The Unity of Forces

A Theory of Universal Equivalence

By

Manoranjan Ghoshal

Copyright

The Unity of Forces: *A Theory of Universal Equivalence*

This is a work of original theoretical physics and reflects the author's independent research, interpretations, and philosophical viewpoints. While every effort has been made to ensure accuracy, the content is presented as a speculative framework and is not intended as a substitute for experimental validation or peer-reviewed consensus.

For permissions or inquiries, contact:

ghoshal.manoranjan@gmail.com

Book Description

What if the four fundamental forces of nature—gravity, electromagnetism, the strong force, and the weak force—are not separate at all, but different faces of a single universal force?

In ***The Unity of Forces:*** *A Theory of Universal Equivalence*, physicist Manoranjan Ghoshal presents a groundbreaking theoretical framework that redefines the way we understand the natural world. Building on his Unit Particle Theory, Ghoshal proposes that all physical forces emerge from the intrinsic behaviour and configurations of a single, indivisible entity—the Unit Particle.

This bold and visionary work explores:

1. A unified origin of all known forces in nature
2. How force arises from the density, motion, and interaction of Unit Particles
3. A reinterpretation of gravity without spacetime curvature
4. New insights into quantum phenomena and cosmic structure
5. A challenge to the Standard Model and General Relativity

Bridging physics, philosophy, and logic, *The Unity of Forces* invites readers to rethink the fabric of the universe through the lens of true material unity.

Acknowledgement

This work would not have been possible without the quiet strength of introspection, the courage to challenge established thought, and the enduring belief in the unity of nature.

I extend my gratitude to those thinkers, both historical and contemporary, who dared to ask "what if" and laid the intellectual foundations on which new theories must be built. I also thank the readers and fellow inquirers who seek truth not only in established facts but in the uncharted possibilities of the universe.

To those who supported my curiosity and encouraged the pursuit of a deeper understanding—your faith means more than words can express.

— *Manoranjan Ghoshal*

Book outline

Preface

***The Unity of Forces:** A Theory of Universal Equivalence*

In the pursuit of understanding the universe, physics has long attempted to identify and categorize the fundamental forces that govern reality. From Newton's gravitational law to the formulation of the electromagnetic spectrum, from quantum field theory to general relativity, each era of science has offered refined, yet fragmented, explanations of the forces at play in nature.

But I now believe that this fragmentation is an illusion—a product of limited perception rather than ultimate truth.

This book is born from a realization that followed my earlier work on Unit Particle Theory: that all known forces—gravitational, electromagnetic, strong nuclear, and weak nuclear—are not separate phenomena, but rather different manifestations of a single, unified force. A force that arises naturally from the properties and configurations of the most elementary entity in existence—the **Unit Particle**.

The journey that led to this understanding has not been conventional. It did not emerge from particle accelerators or supercomputing simulations, but from logical inquiry, mathematical consistency, and the persistent belief that the universe, in its deepest structure, must be governed by simplicity and unity. This belief, I now hold with conviction, points us to a singular foundation: **the equivalence of all forces in origin**.

In this book, I will present the rationale, theoretical framework, and implications of this unifying concept. I am not only to challenge the prevailing paradigms of physics but to offer a new lens through which the universe might be seen—not as a collection of disconnected forces, but as the elegant expression of one.

May this work inspire new questions, bold explorations, and a renewed quest for unity in the heart of physics.

Manoranjan Ghoshal

Author and Theorist

2025

To the Reader

This book invites you to explore a bold reimagining of the physical world—a theory that seeks to unify all known forces through the principle of rotational dynamics and the existence of a single, undividable entity: the Unit Particle.

If you've ever questioned how gravity, electromagnetism, and nuclear interactions might share a deeper commonality, or if you've sensed that the frameworks of modern physics, though powerful, leave some puzzles unsolved—then this book is written for you.

You do not need to be a physicist to engage with these ideas. What you need is curiosity, a willingness to consider new perspectives, and an appreciation for the elegance of nature's underlying order. Through intuitive diagrams, logical reasoning, and real-world analogies, I have aimed to make this theory accessible to both scholars and general readers alike.

I encourage you to read not with the intent to agree or disagree immediately, but to understand and reflect. This book is not the final word on the subject—it is an opening statement in a conversation that I hope will continue in classrooms, laboratories, and among thinkers around the world.

Let us together walk the path toward discovering a unified vision of the universe.

Warm regards,

Manoranjan Ghoshal

Dedication

To the seekers of truth—

those who question, imagine, and never stop wondering

about the nature of reality.

And to the Unit Particle—

silent architect of all existence.

— M.G.

Introduction: The Fragmented View of Force

For centuries, science has sought to reveal the hidden architecture of nature—to decipher the patterns that govern the motion of planets, the dance of atoms, and the unseen energy that holds the cosmos together. At the heart of this grand pursuit lies a fundamental question: *What are the forces that shape our universe?*

Modern physics answers this question with four pillars—the so-called fundamental forces: **gravity**, **electromagnetism**, **the strong nuclear force**, and **the weak nuclear force**. Each is treated as distinct, each described by different equations, operating at different scales, and explained through different mechanisms. This fragmentation of natural force has become a defining feature of contemporary scientific thought.

And yet, something in this segmentation feels incomplete.

Despite the power of current models—from Newton's gravity to Maxwell's electromagnetism, from the Standard Model of particle physics to Einstein's spacetime curvature—none of these frameworks successfully unite all forces under a single, consistent theory. The greatest minds of the past century have worked tirelessly toward a *Theory of Everything*, but no final synthesis has been achieved.

In this book, I present a radically different perspective—one rooted not in mathematical abstraction but in physical reality. I propose that **all forces are, in fact, the same force**—arising from the density, arrangement, and interaction of a single, indivisible entity I call the **Unit Particle**.

This idea builds on my earlier work, *Materialism Theory and the Concept of Unit Particles*, where I introduced the concept of the Unit Particle as the fundamental, undividable unit of matter. Every known particle—electron, proton, neutron—is, in this view, a structured collection of whole Unit Particles. The behaviours we label as forces are not separate phenomena, but emergent effects resulting from the configurations and distributions of these fundamental units.

In the chapters that follow, I will explore how gravity, electromagnetism, and the nuclear forces may all be seen as expressions of **a singular universal force**. I will show how this force emerges not from mystical fields or curved geometries, but from the interaction of real, measurable components. I will attempt to demonstrate that the forces we perceive as different are united in origin—and that this unity offers a new and elegant path toward understanding the nature of existence.

This book is not intended as a rejection of science, but as an expansion of its philosophical foundation. It is a call to rethink assumptions, to reconsider the divisions we have drawn between phenomena, and to embrace the possibility that nature may be simpler, and more unified, than we have ever imagined.

— *Manoranjan Ghoshal*

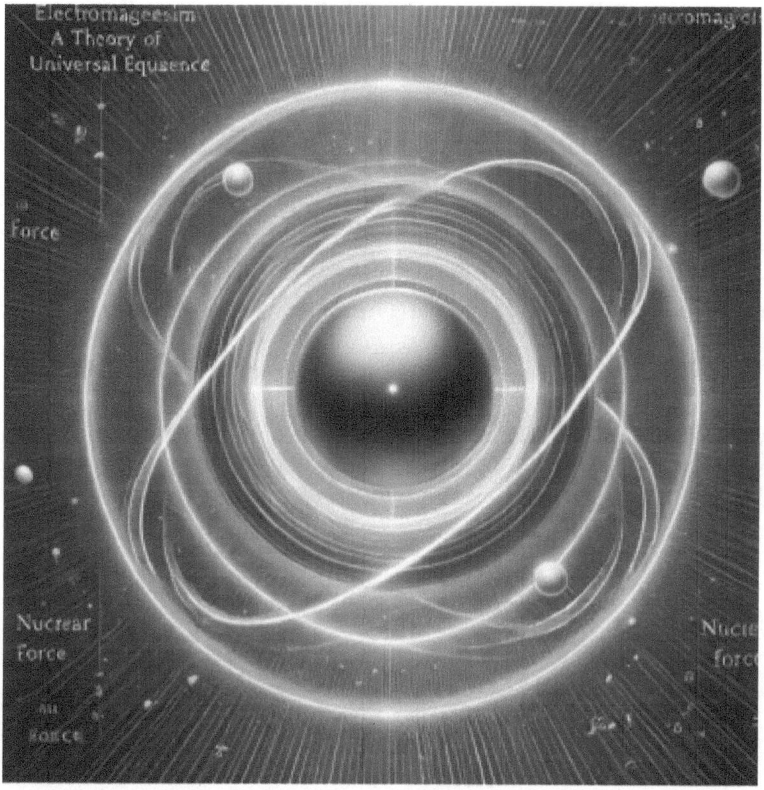
Electromageesim
A Theory of
Universal Equsence
Force
Nucrear
Force
Force

Chapter One: The Historical Journey of Force in physics

The concept of "force" is one of the most foundational in physics, yet also one of the most elusive. From the earliest philosophical musings to the latest high-energy particle experiments, force has been both a measurable effect and a deeply interpreted phenomenon. Understanding its historical evolution helps reveal the assumptions that continue to shape modern science—and why a new perspective is needed.

1.1 Force in Classical Thought

THE EARLIEST RECORDED ideas of force and motion can be traced to ancient Greece. Aristotle believed that force was required to keep objects in motion—a view that persisted for centuries until it was overthrown by the radical simplicity of Newton's First Law.

In the 17th century, **Sir Isaac Newton** revolutionized the understanding of force with his *Principia Mathematica*. He introduced three laws of motion and defined force as the cause of acceleration:

$F = ma$

This simple equation became a cornerstone of classical mechanics, describing how objects interact under the influence of measurable external causes.

But Newton's gravity—an invisible attraction acting at a distance—was still mysterious. He offered no mechanism for it, only a description.

1.2 Electromagnetism: The First Unification

THE NEXT MAJOR LEAP came with **James Clerk Maxwell** in the 19th century. Through elegant mathematics, Maxwell unified electricity and magnetism into a single force: **electromagnetism**. He demonstrated that light itself was an electromagnetic wave—bringing disparate phenomena under one theoretical roof.

This was the first taste of unity in physics. It hinted that nature might not be fragmented, but fundamentally connected.

1.3 The Rise of Quantum and Nuclear Forces

IN THE 20TH CENTURY, the quantum revolution introduced two new forces to the catalogue: the **strong nuclear force**, which holds atomic nuclei together, and the **weak nuclear force**, responsible for radioactive decay.

With the development of the **Standard Model**, these were integrated into a quantum field framework. The strong and weak forces were shown to operate at incredibly small scales and under conditions vastly different from gravity or electromagnetism.

Still, this growing patchwork lacked cohesion. Gravity resisted unification with quantum theory. And while electromagnetism and the weak force were combined under the **electroweak theory**, a full Grand Unified Theory (GUT) remained out of reach.

1.4 General Relativity and the Spacetime Paradigm

WHILE QUANTUM MECHANICS was reshaping the subatomic world, **Albert Einstein** was reimagining gravity. In his **General Theory of Relativity**, gravity was no longer a force in the traditional sense but the result of curved spacetime caused by mass and energy.

This geometric model of gravity stood apart from the force-carrier approach of the Standard Model. As a result, the two great frameworks of physics—relativity and quantum theory—grew in different directions, unable to merge.

1.5 The Fragmentation Problem

MODERN PHYSICS, THEN, remains divided:

1. **Gravity**: Explained by spacetime curvature (General Relativity)
2. **Electromagnetism**: Carried by photons (Quantum

Electrodynamics)
3. **Strong Force**: Carried by gluons (Quantum Chromo dynamics)
4. **Weak Force**: Carried by W and Z bosons

Each force is described by a different mathematical structure, mediated by different particles, and acting on different principles. The result is a brilliant but fractured mosaic.

1.6 A Call for Unity

WHY SHOULD NATURE BE so divided? Why should the universe operate by multiple separate rules at different scales?

Perhaps it doesn't. Perhaps this fragmentation is the result of the way we've chosen to interpret physical interactions, not an intrinsic feature of reality itself.

In the next chapter, we will begin to build the foundation for a **unified view**—one that sees all forces not as separate mechanisms, but as emergent effects of a single, universal interaction rooted in the fundamental nature of matter.

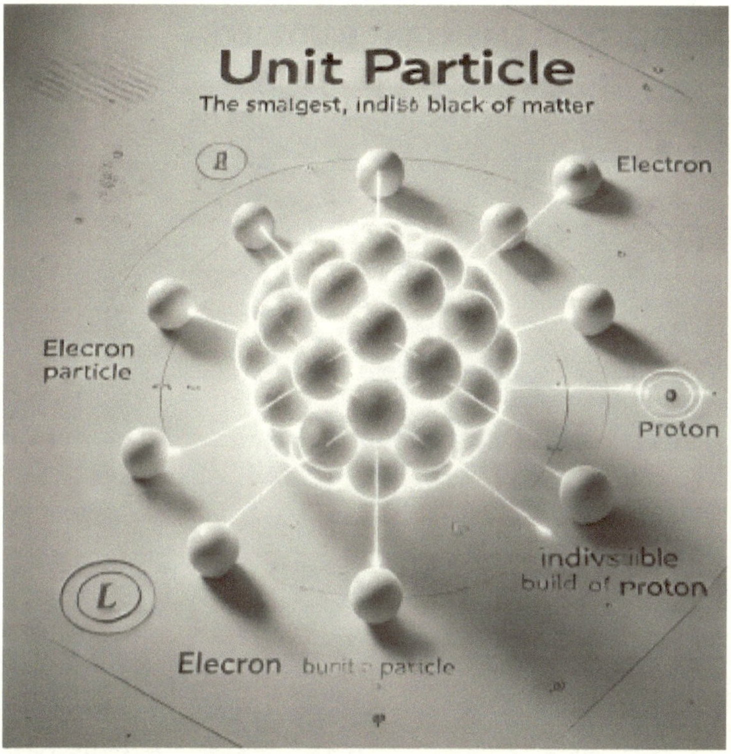
Unit Particle
Electron
Elecron
particle
Proton
build of proton
Elecron

STANDARD MODEL OF ELEMENTARY PARTICLES

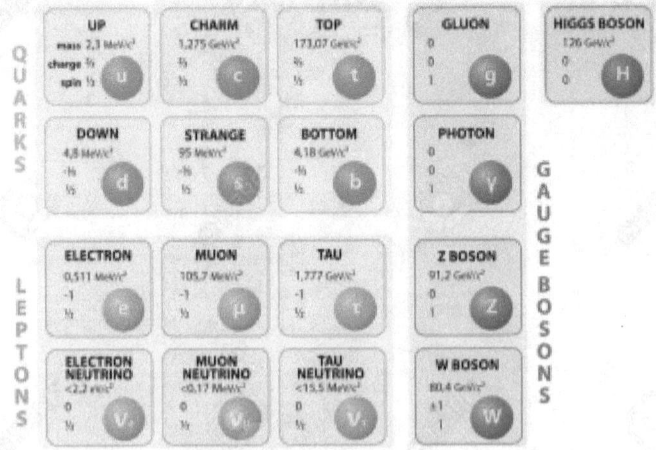

Chapter Two: The Case for a Common Origin

Modern science operates with a deep respect for empirical distinctions. The four known fundamental forces have different strengths, ranges, and behaviours. But beneath these differences lies a question too important to ignore: *Could all these forces arise from a single origin?*

This chapter presents the philosophical and logical foundations for such a unification, exploring how the separation of forces may be an artifact of perception, not a truth of nature.

2.1 Patterns of Unity in Nature

HISTORY HAS REPEATEDLY shown that complexity in science often conceals an underlying simplicity:

1. Maxwell unified electricity and magnetism.
2. Mendeleev organised chemical elements into a periodic table based on atomic patterns.
3. The DNA double helix provided a single code for all life.

These examples reveal a natural tendency toward unification. Nature, it seems, prefers economy. Redundancy is often an illusion caused by lack of perspective.

So, might the same principle apply to force?

2.2 Revisiting the Concept of Force

LET US RETHINK WHAT force actually is. At its most basic, force is a **change in motion**—a shift in energy, a disturbance in configuration, an effect that propagates through matter.

But if all matter is ultimately composed of **Unit Particles**, as proposed in the author's previous work, then force must be a phenomenon that emerges from the *interactions* and *distributions* of these fundamental particles. In other words, **force is a consequence of material configuration**, not an external imposition.

2.3 Similarities Among Forces

DESPITE BEING TREATED as distinct, all four known forces share profound similarities:

1. All operate through **interaction.**
2. All exhibit **range limitations** based on energy and scale.
3. All produce **acceleration** or change in motion.
4. All obey conservation principles: energy, momentum, and sometimes charge.

Could these commonalities point to a shared origin? If so, the differences might be merely **contextual effects**—variations in how a single universal force behaves under different densities, arrangements, and velocities of Unit Particles.

2.4 Forces as Emergent, Not Fundamental

IN THE UNIT PARTICLE framework, the true fundamental entity is not the force itself, but the **material arrangement** that gives rise to it.

Just as temperature emerges from the kinetic energy of molecules, force might emerge from how Unit Particles cluster, move, repel, or attract in structured formations.

This approach eliminates the need for abstract force-carriers like gravitons or gluons. Instead, these could be seen as mathematical representations of real, material interactions—**effects**, not causes.

2.5 The Illusion of Separation

CONSIDER THE FOLLOWING analogy: sound, light, and vibration are perceived through different senses, yet all are forms of wave propagation. Their perceived difference comes from the medium, not the underlying phenomenon.

So too, gravity and electromagnetism may seem unrelated because they operate at different scales and under different material configurations. But fundamentally, they could both be **expressions of the same underlying material dynamics.**

2.6 Philosophical Justification

FROM A PHILOSOPHICAL standpoint, unity is not only desirable—it is **necessary** for true understanding. A theory that explains everything in terms of one indivisible cause is inherently more robust than one requiring multiple unrelated forces.

The concept of a **Universal Force** emerging from **Universal Matter**—the Unit Particles—offers a holistic framework. It aligns with Occam's Razor, respects conservation laws, and speaks to the deep coherence of nature.

2.7 Toward a Unified Theory of Interaction

THIS CHAPTER LAYS THE conceptual groundwork for the rest of the book. From here, we move from philosophy to mechanics—from idea to mechanism. We will explore how gravity, electromagnetism, and the nuclear forces may each be explained as **emergent patterns** arising from Unit Particle behaviour.

In doing so, we aim to not only unify force—but to understand its **true nature** for the first time.

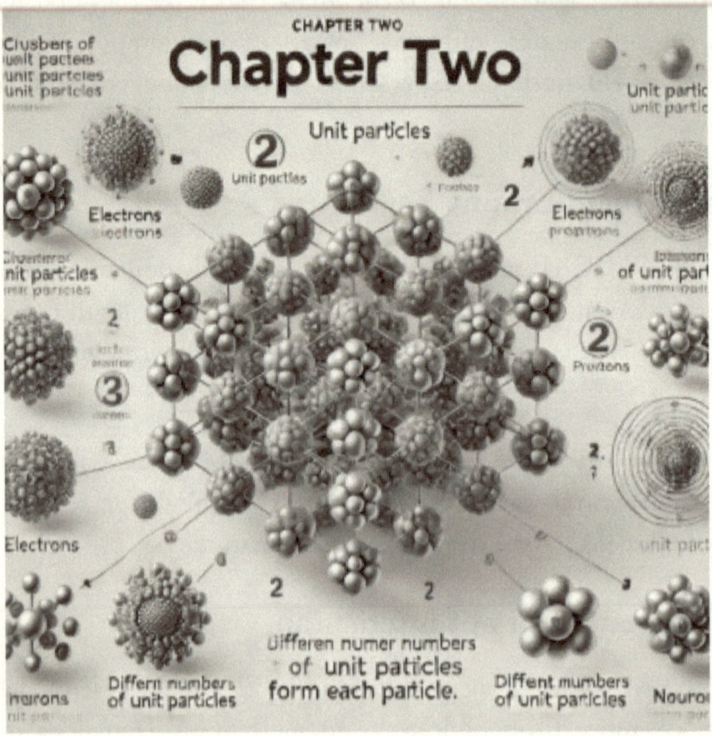
CHAPTER TWO
Chapter Two
Unit particles
Electrons
Electrons
of unit par
Protons
Electrons
Differn numbers
of unit particles
Ulfferen numer numbers
of unit paticles
form each particle.
Diffent mumbers
of unit paricles

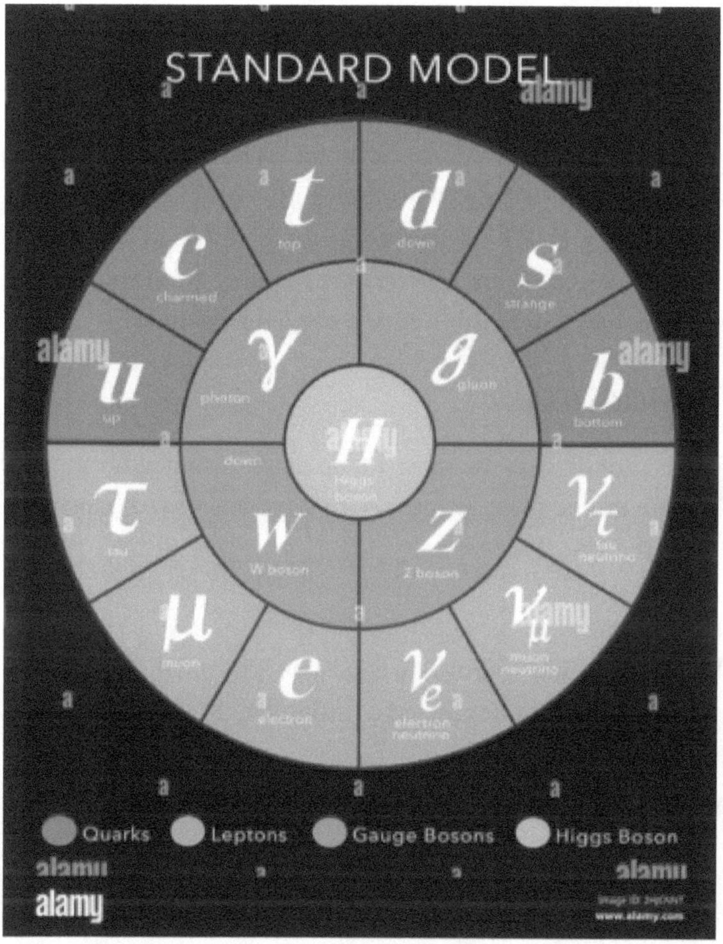
STANDARD MODEL
t
top
d
down
c
charmed
s
strange
u
up
γ
photon
g
gluon
b
bottom
H
τ
W
W boson
Z
Z boson
μ
e
electron
Quarks
Leptons
Gauge Bosons
Higgs Boson

Chapter Three: Unit Particle Theory – A Brief Revisit

Before we delve into how all known forces may be unified through a single mechanism, we must briefly revisit the central idea on which this framework rests: **Unit Particle Theory (UPT)**. It serves as the bedrock of this book's thesis—the claim that all forces are emergent from the arrangement and interactions of one elementary, indivisible entity: the **Unit Particle.**

3.1 The Unit Particle Defined

THE **Unit Particle** is proposed as the most fundamental, irreducible particle in existence. Unlike subatomic particles known to physics—such as electrons, protons, and quarks—the Unit Particle is:

1. **Indivisible**: It cannot be split into smaller parts.
2. **Invariable**: Its mass and volume are fixed and constant.
3. **Indestructible**: It cannot be created or annihilated—only arranged or reconfigured.
4. **Material**: It has real physical presence, not abstract field identity.

This particle is the **atomic essence of matter**, not in the chemical sense, but in the absolute sense. Everything—matter, energy, force—emerges from the configuration and motion of these particles.

3.2 Building the Known Universe

EVERY KNOWN PARTICLE—ELECTRON, proton, neutron, photon—is, under UPT, a structured **conjugation** of multiple Unit Particles.

Let:

1. x = number of Unit Particles in an electron
2. y = number of Unit Particles in a proton, where $y > x$

Each known particle corresponds to an **integer** grouping of Unit Particles. This integer quantization leads to measurable and discrete masses, charges, and behaviours.

In this model:

1. Particles are **material structures.**
2. Forces arise from **interactions** between these structures.
3. Fields are **manifestations** of spatial configurations of Unit Particles.

3.3 Movement and Interaction

WHEN UNIT PARTICLES group together, their **relative arrangement and density** affect their behaviour:

1. **Low-density regions** may correspond to less interaction (e.g., photons).
2. **High-density clusters** create strong binding effects (e.g., nuclear forces).
3. **Directional movement** of Unit Particles gives rise to wave phenomena and energy.

Thus, **motion** of Unit Particles explains the origin of momentum, while **imbalances** in distribution generate forces.

3.4 Replacing the Standard Model Framework

THE STANDARD MODEL depends on:

1. Fundamental fields (electromagnetic, strong, weak)
2. Force-carrier particles (photons, gluons, W/Z bosons)
3. Probabilistic quantum behaviour

In contrast, UPT offers:

1. A **single substance** (Unit Particle) as the basis for all physical reality

2. **Deterministic structure** defined by configuration and motion
3. Force as **emergent material interaction**, not field abstraction

3.5 Ontological Simplicity

ONE OF THE GREATEST strengths of Unit Particle Theory is its **philosophical and ontological simplicity**:

1. No arbitrary field constructs
2. No unexplained particle zoo
3. No abstract carriers or virtual entities

It posits **one origin** for all structure, motion, and force—a singular, universal building block whose permutations yield the rich complexity of the cosmos.

3.6 Preparing for Unified Force Interpretation

WITH THIS REFRESHER, we are now equipped to approach the central aim of this book: the **unification of all forces** through the lens of Unit Particle Theory. In the next chapter, we will explore how **gravity**, traditionally seen as the outlier among the four forces, can be reinterpreted within this framework—not as a curvature of spacetime, but as a **density effect of real particles.**

Before of that you know what is field?

3.4a Understanding the Concept of a Field

IN TRADITIONAL PHYSICS, a **field** is defined as a region of space in which a physical quantity—such as force, energy, or potential—can be measured at every point. Fields are used to explain how particles interact across distance without direct contact. For example:

1. A **gravitational field** assigns a force per unit mass to each point in space due to a mass source.
2. An **electric field** does the same for electric charge.
3. A **magnetic field** arises from moving electric charges and affects

other moving charges.

These fields are often treated as **fundamental entities** in physics, with their own independent existence—particularly in quantum field theory (QFT), where particles themselves are considered as *excitations in underlying fields*.

However, this approach introduces complexity. Fields become abstract constructs that must fill all of space, carry energy, and interact with each other in highly mathematical, often non-visualisable ways.

3.4b Fields Reimagined in Unit Particle Theory

IN **Unit Particle Theory**, a field is **not a separate entity**. Rather, it is:

A **perceptual or mathematical abstraction** of how Unit Particles are **arranged and move** within space.

According to UPT:

1. There are no immaterial fields stretched across space.
2. What we interpret as a field is actually the **effect** produced by a **distribution** or **gradient** of Unit Particles.
3. When the density or movement of Unit Particles varies across space, this creates **force gradients**—and we call this a field.

So, instead of a **gravitational field** pulling objects, UPT describes:

A **high-density region of Unit Particles** exerting directional influence on lower-density regions nearby.

The "field" is thus a way to map and describe these interactions—**not an actual thing**, but a useful way to talk about **how Unit Particles influence each other across space**.

3.4a Understanding the Concept of a Field

IN TRADITIONAL PHYSICS, a **field** is a region of space where a physical quantity, such as force or energy, has a value at every point. Fields are used to explain how objects can exert forces on each other over distances. For example:

1. **Gravitational Field**: Describes the gravitational force exerted by a mass at various points in space.
2. **Electric Field**: Represents the force experienced by a charged particle due to other charges in its vicinity.
3. **Magnetic Field**: Depicts the influence exerted by magnetic materials or moving charges.

These fields are often visualized as invisible lines permeating space, indicating the direction and strength of the force.

3.4b Fields Reimagined in Unit Particle Theory

IN **Unit Particle Theory**, the concept of a field is reinterpreted. Instead of being fundamental entities that exist independently, fields are seen as emergent properties resulting from the distribution and movement of **Unit Particles**—the indivisible building blocks of matter.

Analogy: The Crowd Movement

Imagine a densely packed crowd of people:

1. **High-Density Region**: In areas where people are closely packed, movement is restricted. If someone tries to move, they experience resistance due to the surrounding crowd.
2. **Low-Density Region**: In areas where people are sparse, movement is easier, with little to no resistance.

In this analogy, the "resistance" one feels moving through the crowd isn't due to an invisible force field but arises from the physical presence and arrangement of people. Similarly, in UPT, what we perceive as a field is actually the result of the density and arrangement of Unit Particles in a given region.

Key Points in UPT:

1. **No Separate Fields**: Fields don't exist as separate physical entities but are manifestations of Unit Particle configurations.
2. **Emergent Forces**: Forces arise due to variations in Unit Particle densities. For instance, gravitational attraction is a result of differences in Unit Particle concentrations between two regions.
3. **Directional Influence**: The movement or flow of Unit Particles can create directional influences, analogous to how wind emerges from air pressure differences.

Visual Representation

TO AID IN VISUALIZING this concept, consider the following diagrams

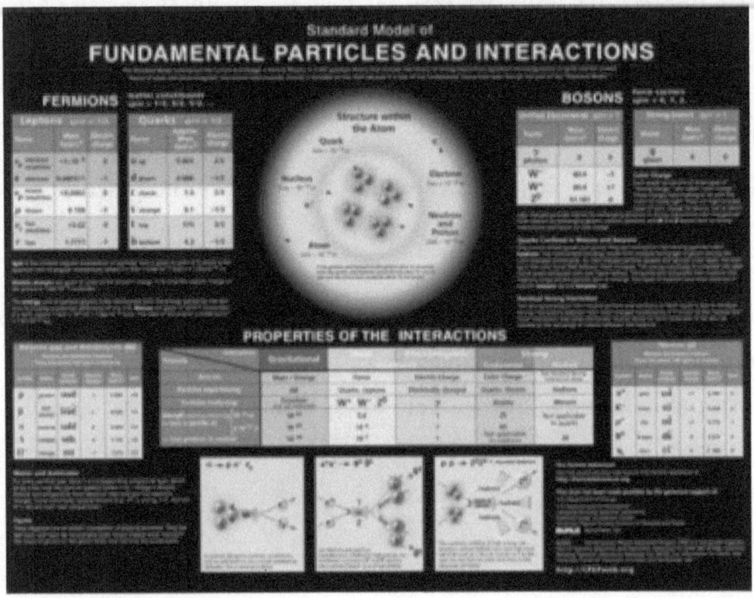

In this diagram:

1. **Left Side**: Depicts a uniform distribution of Unit Particles, resulting in no emergent force.
2. **Right Side**: Shows a gradient in Unit Particle density, leading to an emergent force from the high-density region to the low-density region.

This visualization emphasizes that what we traditionally interpret as a "field" is, in UPT, the outcome of Unit Particle distributions.

By reimagining fields in this manner, Unit Particle Theory offers a more tangible and unified explanation for the origin of forces, grounding them in the physical interactions of fundamental particles rather than abstract entities.

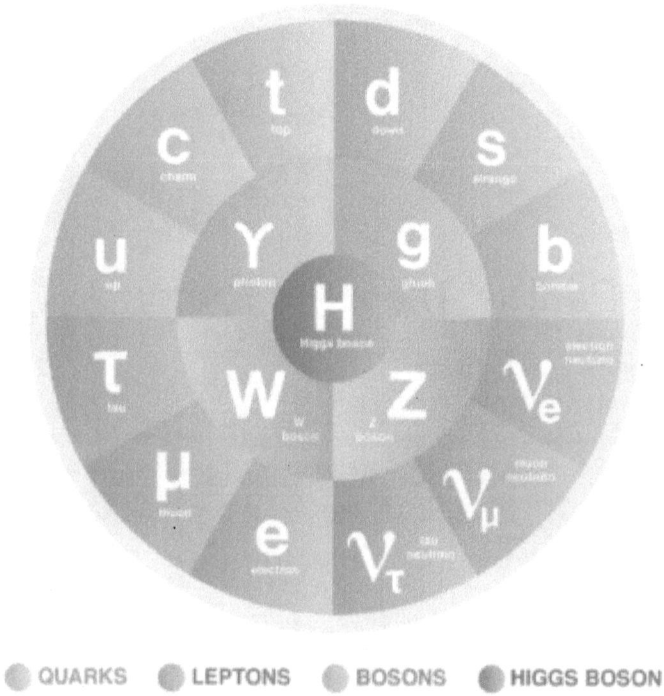
t
d
c
s
u
γ
g
b
H
τ
W
Z
ν_e
μ
ν_μ
e
ν_τ
QUARKS
LEPTONS
BOSONS
HIGGS BOSON

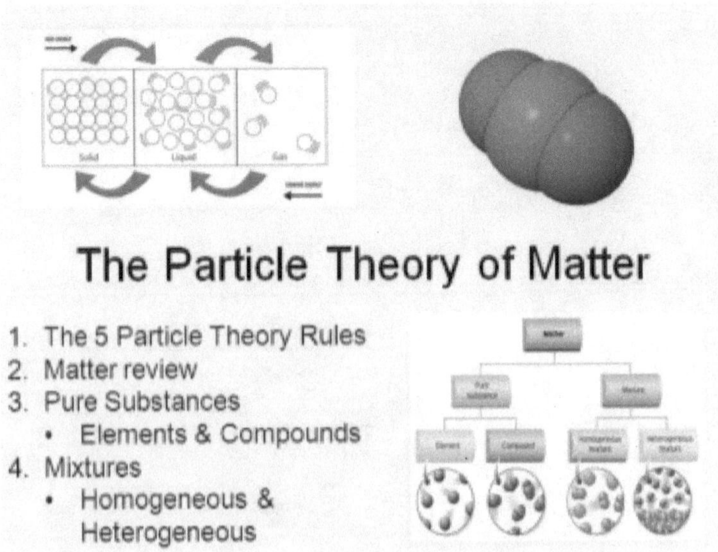
The Particle Theory of Matter
1. The 5 Particle Theory Rules
2. Matter review
3. Pure Substances
• Elements & Compounds
4. Mixtures
• Homogeneous & Heterogeneous

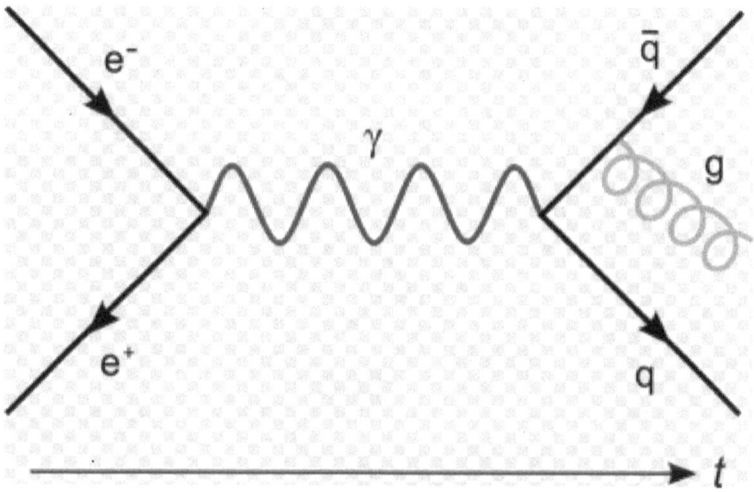
e⁻
e⁺
γ
q̄
g
q
t

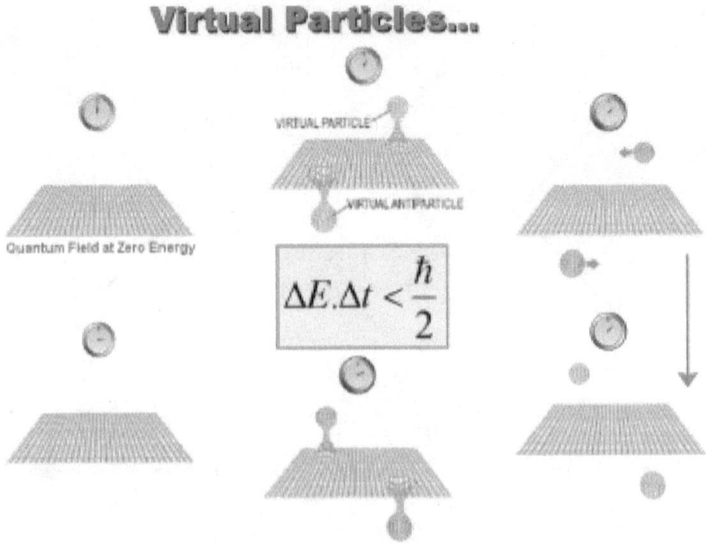
Virtual Particles...
VIRTUAL PARTICLE
VIRTUAL ANTIPARTICLE
Quantum Field at Zero Energy
$\Delta E.\Delta t < \frac{\hbar}{2}$

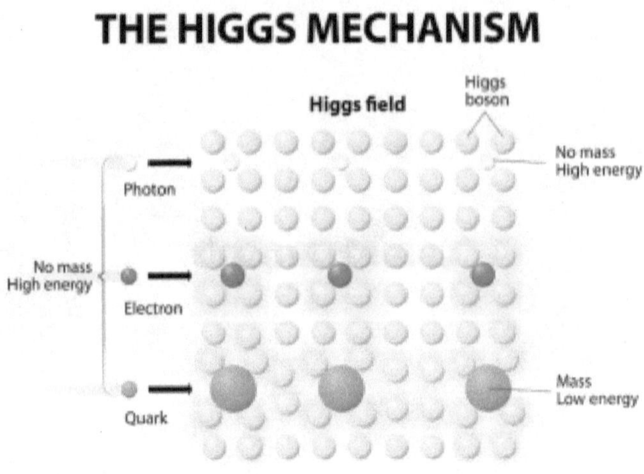
THE HIGGS MECHANISM
Higgs field
Higgs boson
No mass
High energy
Photon
No mass
High energy
Electron
Quark
Mass
Low energy

Particle Theory of Matter
SNC1D - Chemistry Unit

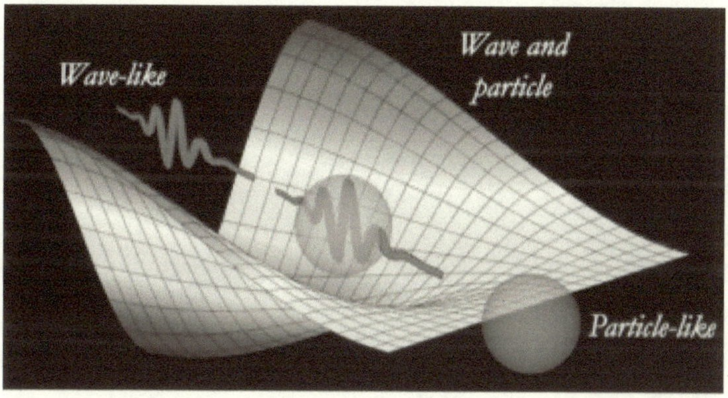
Wave-like
Wave and particle
Particle-like

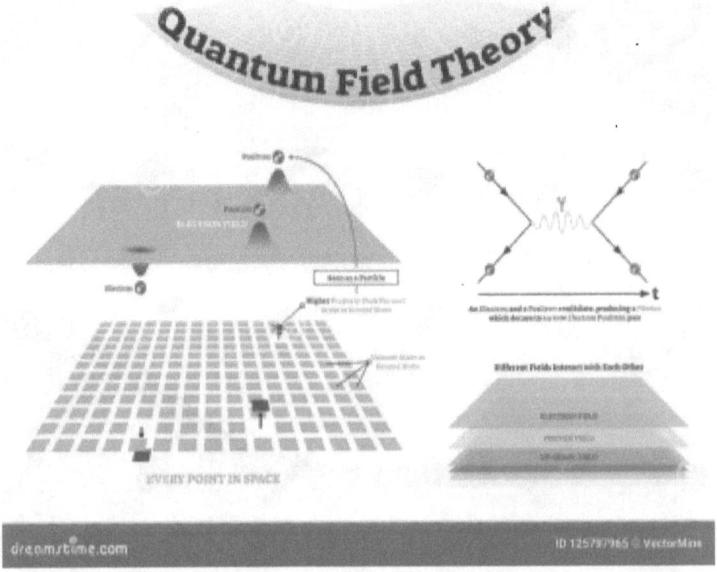
Quantum Field Theory
EVERY POINT IN SPACE
dreamstime.com
ID 125797965 © VectorMine

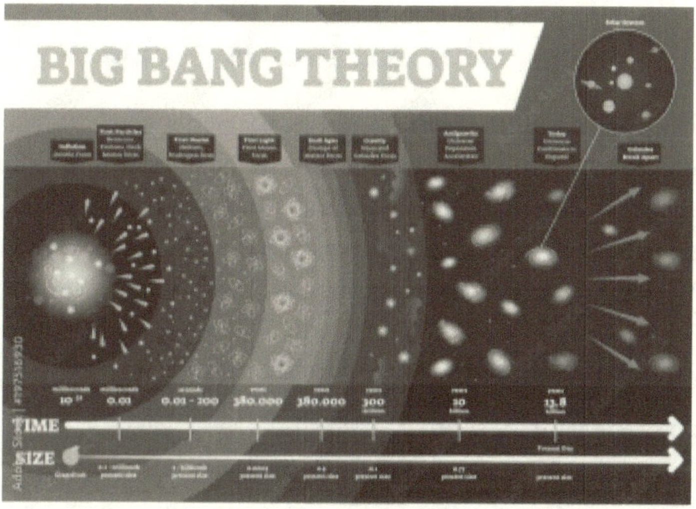
BIG BANG THEORY
0.01
0.01 - 200
380.000
380.000
300
20
13.8
TIME
SIZE

Chapter three point five: The Rotational Basis of All Fields

IN CONVENTIONAL PHYSICS, each field—gravitational, electric, magnetic—has been historically treated as distinct, governed by different rules and force carriers. But within **Unit Particle Theory (UPT)**, all fields emerge from a single, unifying behaviour: the **rotational configuration and dynamics of Unit Particles around a central point.**

3.5.1 Rotation as the Primary Mechanism

ROTATION IS EVERYWHERE in nature:

1. **Planets and stars** rotate on their axes and orbit centers of gravity.
2. **Electrons** in atoms orbit nuclei in discrete energy levels.
3. **Magnets** induce spin alignment in ferromagnetic materials.
4. **Electric fields** produce current flow, which in turn induces magnetic fields—a rotating feedback loop.

These observations suggest a **universal rotational pattern** underpinning all fields and forces.

3.5.2 The Unit Particle Rotation Model

IN UPT, PARTICLES LIKE electrons, protons, and even atomic nuclei are **structured arrangements of Unit Particles**, bound in rotational motion:

1. This rotation creates **internal symmetry** and **stability**.
2. The **axis of rotation** determines the direction and polarity of resulting effects (e.g. magnetic poles).
3. Rotation causes **density gradients** and **momentum flows**, which we interpret as fields.

3.5.3 Field Manifestations from Rotation

LET'S BREAK THIS DOWN across known fields:

Gravitational Field

1. A mass rotates (planet, star, etc.) → Unit Particle motion forms a **dense rotating shell.**
2. Nearby masses are drawn toward **equilibrium with this density gradient**—perceived as gravity.

Electric Field

1. A charged particle has internal rotation (structural asymmetry).
2. This rotation establishes **outward or inward Unit Particle displacement**, depending on polarity—hence positive or negative charge.
3. Static fields are radial; moving charges generate magnetic fields too.

Magnetic Field

1. Created by the **rotational alignment** of Unit Particles within materials.
2. In ferromagnetism, applying a magnet induces **rotational reconfiguration** of internal Unit Particles in ordinary iron.
3. One pole thrusts out, the other pulls in—creating a **closed-loop circuit** of particle flow.

3.5.4 Contact Magnetization and Closed Circuits

CONTACT-BASED MAGNETIZATION (e.g., rubbing a magnet on iron) shows how rotation can **restructure material internally**:

1. The **magnet's rotational field** imposes a new orientation on the iron's Unit Particles.
2. This reorientation creates **a unified flow loop**—one pole acts as a **particle source**, the other as a **receiver.**

Thus, **magnetism** is not an abstract field but a **rotationally induced configuration** of Unit Particles—**a closed circuit of directed motion.**

3.5.5 The Universal Field Principle

FROM THIS VIEW, **all fields—gravitational, electric, magnetic—are different faces of one phenomenon**:

The structured rotational behaviour of Unit Particles around a central point.

1. **Direction** of rotation determines polarity.
2. **Density** of rotating Unit Particles determines force strength.
3. **Speed** and **asymmetry** produce dynamic field effects like induction or wave propagation.

This principle lays the foundation for truly **unified physics**—where forces are not separate, but structurally derived.

Chapter Three Point Six: The Nature of Fields—Density, Rotation, and Dual Force Dynamics

3.6.1 Introduction: What Is a Field in UPT?

IN CLASSICAL PHYSICS, a **field** is an abstract quantity assigned to every point in space—electric, magnetic, gravitational. But in **Unit Particle Theory (UPT)**, a field has a **material basis**:

A field is a region of space structured by the **rotational organization** and **density gradient** of Unit Particles.

Wherever particles or masses exist, they generate fields—not through transmission of force, but through their **dynamic internal rotation** and resulting influence on surrounding Unit Particles.

3.6.2 Central Density and Radial Decline

EVERY FIELD IN UPT shares a common density structure:

1. **Highest Unit Particle density is at the centre**, where mass and rotational activity are most concentrated.
2. As distance increases, **rotational influence spreads**, causing **density to fall off gradually** with radius.
3. This forms a **smooth density gradient** from the core outward, establishing the fundamental shape of a field.

This structure mirrors natural phenomena:

1. Planetary atmospheres are densest near the surface.
2. Electric fields decrease with distance squared.
3. Magnetic flux lines expand out and loop back, creating symmetry in their distribution.

3.6.3 Angular Momentum Governs Field Balance

THIS ROTATIONAL FIELD maintains strict **conservation of angular momentum**:

1. Each layer of Unit Particles moves in harmony, adjusting **velocity and radius** to maintain the same angular momentum per shell.
2. The **rotation period** remains balanced, keeping the system **stable and coherent.**
3. Any interaction with other particles or fields must obey this rotational conservation.

The result is a **self-sustaining field**, always seeking dynamic equilibrium through its particle motion.

3.6.4 Attraction and Repulsion from Density Gradient

THIS IS THE CORE INSIGHT:

Both attraction and repulsion emerge naturally from how another body interacts with the field's Unit Particle density.

1. **Attraction** occurs when the external object has **lower density** or complementary rotational alignment—it moves **inward** to restore balance.
2. **Repulsion** arises when the object has **higher density** or conflicting rotational configuration—it is pushed **outward** to preserve angular harmony.

This duality explains how:

1. **Electromagnetic fields** exhibit both poles and forces.
2. **Gravity**, though primarily attractive, may be complemented by repulsive effects in exotic configurations (as in cosmic inflation or dark energy zones).
3. **Atomic repulsion** at very small distances arises despite gravitational

attraction—because internal rotational density creates an exclusion zone.

3.6.5 Field Boundaries and Zones

EVERY FIELD HAS THREE primary zones:

1. **Core Zone** – Maximal Unit Particle density and fastest internal rotation.
2. **Transition Zone** – Rapid density drop-off; interaction zone for attraction/repulsion forces.
3. **Outer Zone** – Sparse Unit Particle presence; field fades but rotational influence continues subtly.

Field strength and behaviour are dictated by:

1. Total Unit Particle mass.
2. Rotational speed and structure.
3. Interference or merging with nearby fields.

3.6.6 Conclusion: A Material, Dynamic, Dual-Force Field

FIELDS IN UPT ARE:

1. **Real and structured**, not abstract.
2. Defined by **rotational flow** and **density variation**.
3. Capable of producing **both attraction and repulsion**, depending on configuration and angular momentum dynamics.

This understanding reshapes our concept of fields: not invisible carriers of force, but **dynamic, physical environments** created by the collective rotation of fundamental, indivisible Unit Particles.

Chapter Four: Gravity as a Rotational Density Effect

4.1 Gravity: Beyond Curvature and Force

TRADITIONAL THEORIES offer two dominant views on gravity:

1. **Newtonian Gravity** treats it as a force acting at a distance between two masses.
2. **Einstein's General Relativity** interprets it as the curvature of spacetime caused by mass and energy.

However, neither view offers a **material basis** for gravity—what exactly exerts the force or how curvature physically forms remains unresolved.

Unit Particle Theory (UPT) offers an alternative explanation:

Gravity is the result of rotational density gradients of Unit Particles.

This model provides a tangible, physical cause—rooted in the intrinsic rotation of matter at all scales.

4.2 Rotation and the Formation of Mass

ALL KNOWN MASSIVE BODIES—FROM electrons to galaxies—exhibit **rotation**:

1. **Electrons** are arranged rotating Unit Particles.
2. **Protons and neutrons** are denser arrangements with complex internal rotation.
3. **Planets and stars** rotate around their axes.

These rotations create **internal particle currents**, establishing a structured flow of Unit Particles that surrounds the mass.

4.3 Gravity as an Emergent Gradient

IN UPT:

1. Rotating Unit Particles in a body create a **spherical field of rotational motion**.
2. This motion causes **compression** of Unit Particles near the centre and **rarefaction** farther out.
3. The result: a **gradient in Unit Particle density**, strongest near the mass.

This gradient is what we perceive as **gravitational attraction**:

Objects are drawn **into regions of higher rotational density**—not by pull, but by rebalancing through structural interaction.

4.4 Visual Analogy: Water Funnel

IMAGINE A **drain funnel** in water:

1. Water near the drain rotates rapidly, pulling nearby water inward.
2. Objects dropped into the funnel spiral inward—not due to suction, but because of the **density and rotational velocity gradient**.

Gravity in UPT behaves similarly:

1. A massive object causes **Unit Particle swirl**.
2. Nearby objects drift inward along this gradient, **seeking structural equilibrium** in the rotational field.

4.5 The Role of Axis and Polarity

UPT EXPLAINS WHY:

1. **Planets** always spin around an axis.
2. **Orbits** form naturally—due to **symmetrical rotation patterns** of the central mass and its satellites.

3. **Spin orientation** of astronomical objects correlates with gravitational interactions.

Rotational alignment determines:

1. The **flow direction** of Unit Particles.
2. The **plane of orbital movement**.
3. **Field strength** and **range**.

4.6 Implications and Reinterpretations

WITH GRAVITY AS A **rotational density effect**, we can reinterpret several phenomena:

1. **Tidal forces** result from overlapping rotational density fields of Earth and Moon.
2. **Black holes** are not infinite-density points, but **closed-loop collapses of Unit Particle rotation**, forming a maximum density vortex.
3. **Gravitational lensing** is **rotational refraction**—the bending of light caused by its movement through asymmetric Unit Particle density fields.

4.7 Rethinking Inertia and Freefall

IN THIS VIEW:

1. **Inertia** is the **stability of internal Unit Particle rotation**. A body resists acceleration when its rotational configuration is undisturbed.
2. **Freefall** is not motion in curved spacetime, but **the body's attempt to structurally equalize with an external density field.**

Thus, gravity, inertia, and mass are **all properties of rotational Unit Particle arrangements**.

4.8 Summary: Gravity as Natural Structure

UPT OFFERS A BOLD REINTERPRETATION:

Gravity is not a pulling force or geometric illusion—it is the natural outcome of organised rotational density within the material substrate of the universe.

Every particle's rotation contributes to the structure of space around it. Gravity is the dance of these structures, moving and responding in search of energetic balance.

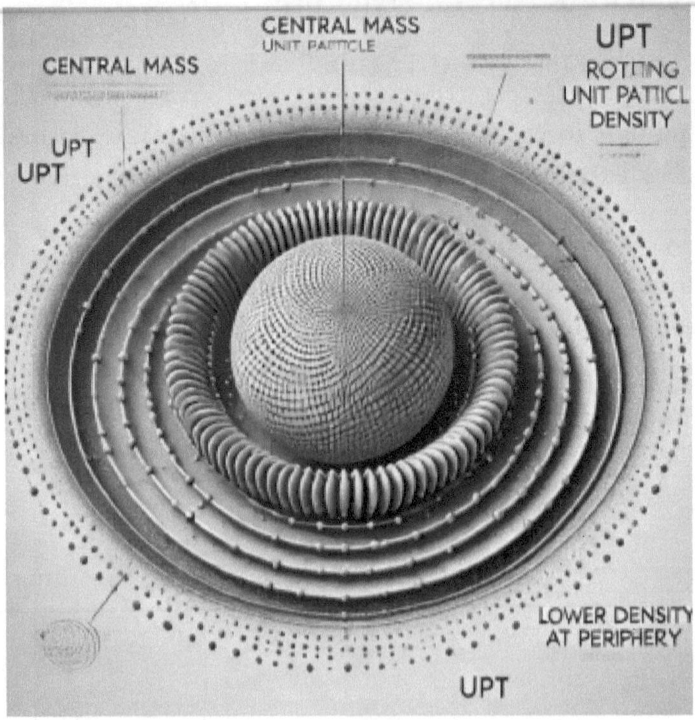
CENTRAL MASS
UNIT PARTICLE
UPT
CENTRAL MASS
ROTTING
UNIT PATTICL
DENSITY
UPT
UPT
LOWER DENSITY
AT PERIPHERY
UPT

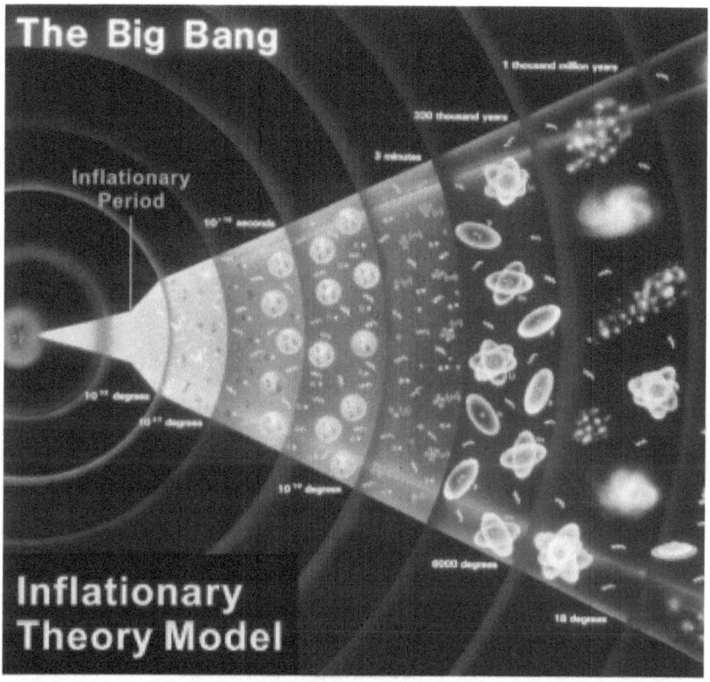

Chapter Five: Electromagnetism as Rotational Displacement

5.1 Introduction: A New View of Electromagnetism

IN CLASSICAL PHYSICS, **electromagnetism** is one of the four fundamental forces, uniting electric and magnetic fields. But UPT redefines it:

Electromagnetism is the result of rotational displacement and alignment of Unit Particles.

Rather than treating electric and magnetic phenomena as separate, UPT sees them as two expressions of the **same underlying rotation-based field structure.**

5.2 Electric Fields as Asymmetric Rotational Flow

AN **electric field** arises when there is a **directional imbalance** in Unit Particle rotation:

1. When a body (like a charged particle) has a net **outward or inward displacement** of rotational alignment, it produces an electric field.
2. The **density gradient** and direction of rotation cause a **structural shift** in nearby Unit Particles.

Positive Charge: A region where Unit Particles rotate outward, causing adjacent particles to be pulled inward (attraction for negative charge).

Negative Charge: A region where Unit Particles rotate inward, creating compression and outward push (repelling like charges).

Electric field lines in UPT aren't mere mathematical arrows—they represent **real movement paths** of rotational influence in the Unit Particle substrate.

5.3 Magnetic Fields as Circular Rotation Alignment

WHEN UNIT PARTICLES rotate **uniformly in closed loops**, a **magnetic field** is formed:

1. This occurs when **charged particles move**—such as in electric current or in naturally magnetized materials.
2. The motion realigns the surrounding Unit Particles into **rotating vortices.**

This alignment:

1. Forms a **dipole**: two poles of concentrated rotational activity.
2. Causes lines of force to emerge from one end (north) and return through the other (south).

Magnetism in UPT is not an effect of invisible force carriers—it is a **visible alignment of real rotating elements** of matter.

5.4 Electromagnetic Induction: A Rotational Exchange

ELECTROMAGNETIC INDUCTION—LIKE Faraday's law—becomes elegantly simple in UPT:

1. When you **rotate a magnetic field**, you are **reorganizing Unit Particle rotation.**

1. This **disrupts or induces rotation** in neighboring Unit Particles, which manifests as **current** in a conductor.

The key is the **transfer of rotational momentum**—not the movement of invisible waves, but actual mechanical-like rotation at microscopic scales.

5.5 Electric Currents and Fields as Organised Flow

ELECTRIC CURRENT IS reinterpreted:

1. Not simply electron flow, but **coordinated directional rotation** of Unit Particles within a conductor.
2. Conductive materials allow **easy reorientation** of Unit Particles due to their loose structural bonding.
3. Insulators resist such reorganisation, hence no field is transferred.

Electric potential (voltage) is then:

A **difference in rotational density**—a form of "rotational pressure"—seeking balance.

5.6 Why Magnetism Can Be Induced but Not Separated

MAGNETISM ALWAYS FORMS dipoles in UPT because:

1. Rotational flow is circular and must **return to its origin.**
2. You cannot isolate one pole without breaking the symmetry of rotation.

Hence, **no magnetic monopole** has been found—not because they don't exist, but because the structure of rotational Unit Particle alignment **naturally forbids it.**

5.7 Implications of UPT for Electromagnetic Technologies

THIS NEW VIEW EXPLAINS:

1. **Magnetization** of iron: Internal realignment of rotating Unit Particles.
2. **Electric motors**: Force arises from controlled rotation of internal fields.
3. **Wireless communication**: Transmission involves rotational modulation of surrounding Unit Particle fields.

We're not "sending waves" through emptiness—we're modulating **real rotating structures** that already pervade space.

5.8 Summary: A Unified, Mechanical Basis for Electromagnetism

IN UPT:

1. Electric and magnetic fields are two aspects of **rotational displacement and density control**.
2. Charge is an expression of **asymmetric Unit Particle spin alignment**.
3. Magnetism is the result of **closed-loop rotation**.
4. All electromagnetic interaction is built from the **motion and alignment of real, indivisible Unit Particles**.

This gives us a **material, unified, and comprehensible** explanation for a once-abstract force.

Chapter Six: The Strong and Weak Nuclear Forces as Compact Rotational Networks

6.1 Introduction: Beyond the Illusion of Separate Forces

IN THE CONVENTIONAL framework of the Standard Model, the **strong** and **weak nuclear forces** are treated as distinct fundamental interactions. The **strong force** binds protons and neutrons within atomic nuclei, while the **weak force** governs radioactive decay and particle transformation.

But from the **Unit Particle Theory (UPT)** viewpoint, these so-called forces are **not fundamental.** They are **manifestations of rotational arrangements, densities, and transitions of Unit Particles** within subatomic systems. Instead of invisible carriers or quantum fields, UPT presents **concrete, rotation-based mechanisms** built from real physical structures.

6.2 The Strong Nuclear Force: Rotational Interlock of Unit Particles

THE STRONG FORCE IS observed to be:

1. The most powerful of the four known forces,
2. Active only at subatomic distances (≈1 fm),
3. Responsible for overcoming the repulsion between positively charged protons in a nucleus.

UPT Interpretation:

THE STRONG INTERACTION arises when **Unit Particles form a compact, interlocked rotational lattice:**

1. Within protons and neutrons, Unit Particles rotate in tightly synchronized configurations.
2. When nucleons come extremely close, the **outer layers of Unit Particles intermesh**, forming a **shared rotational bond.**

3. This bond is **highly stabilizing**, as the rotational direction of each particle supports the next, forming a **coherent rotational network.**

Key Mechanism: Rotational Binding Pressure

1. The **higher the density and synchrony** of Unit Particles, the **greater the internal tension** resisting separation.
2. This rotational pressure forms a **self-reinforcing binding mechanism**—the origin of the strong nuclear force.

THIS EXPLAINS:

1. **Why the strong force only works at short ranges**: Once Unit Particles move beyond their rotational influence (a few femtometers), the interlock fails.
2. **Why it's so strong**: It is built from the **core physical rotation of the universe's most basic building blocks**, producing maximal stability.

6.3 Nucleon Structures and Isotopic Behaviour

DIFFERENT ELEMENTS and isotopes are characterized by:

1. The number of **protons and neutrons,**
2. Their **binding energy,**
3. Their **stability or radioactivity.**

UPT Insight:

1. Protons and neutrons differ in **rotational orientation and density** of Unit Particles.
2. The **ratio of rotational overlap** between nucleons dictates stability.
3. Some isotopes have **imbalanced rotational energy**, which leads to **spontaneous decay**—not because of external forces, but due to internal **rotational disharmony.**

THUS, THE PERIODIC table is a **map of rotational stability**, not just atomic structure.

6.4 The Weak Nuclear Force: Rotational Reconfiguration and Particle Transition

THE WEAK FORCE IS RESPONSIBLE for:

1. Transforming one particle into another (e.g., neutron to proton),
2. Enabling **beta decay**,
3. Changing **flavors of quarks** (in the Standard Model).

UPT Explanation:

RATHER THAN RELYING on hypothetical W and Z bosons, UPT sees weak interactions as:

Internal rearrangements in the rotational patterns of Unit Particles.

1. When the **rotational balance** of a neutron is disturbed—due to external energy or internal instability—the structure **reconfigures**.
2. This shift results in a **proton**, plus an **ejection of excess rotational energy**, expelled as **an electron and a neutrino** (two rotational packets with net spin and motion).

This reconfiguration:

1. Preserves **rotational conservation laws** (angular momentum and mass),
2. Explains **why decay products appear suddenly and at high speed**—they are surplus rotational modes seeking equilibrium.

6.5 Particle Decay in UPT

ALL DECAY PROCESSES—ALPHA, beta, gamma—are **manifestations of unstable Unit Particle structures**:

1. **Alpha decay**: loss of a stable rotational unit (a helium nucleus),
2. **Beta decay**: reorganisation of internal rotations within a neutron or nucleus,
3. **Gamma radiation**: the release of residual rotational energy without mass displacement.

In every case, UPT maintains that **rotational density seeks stability**. Decay is not a mystery but an **inevitable shift toward balance** in a densely packed rotational system.

6.6 Gluons and Bosons: No Longer Needed

IN THE STANDARD MODEL:

1. **Gluons** hold quarks together,
2. **W and Z bosons** mediate weak interactions.

In UPT:

1. **No intermediate particles are required.**
2. The **unit particle network itself creates and sustains all interactions** through:
    - Density-driven interlocking,
    - Internal reconfiguration of rotation,
    - Energy shifts between layered rotational states.

THIS **greatly simplifies the model of matter**, unifying interactions under a **single, materialistic logic**.

6.7 Summary: A Rotation-Based Origin of Nuclear Interactions

Concept	Standard Model	UPT Interpretation
Strong Force	Gluon-mediated force	Interlocking rotation of Unit Particles
Weak Force	Boson-mediated flavor change	Rotational reconfiguration of Unit Particle clusters
Particle Stability	Binding energy of nucleons	Harmonized internal rotation of Unit Particle networks
Decay Processes	Probabilistic particle transformation	Natural release of imbalance in rotational density

THROUGH UPT, THE NUCLEAR forces are no longer isolated mysteries but part of a **coherent, mechanical structure** rooted in the most fundamental feature of matter: the **rotation and density** of indivisible Unit Particles.

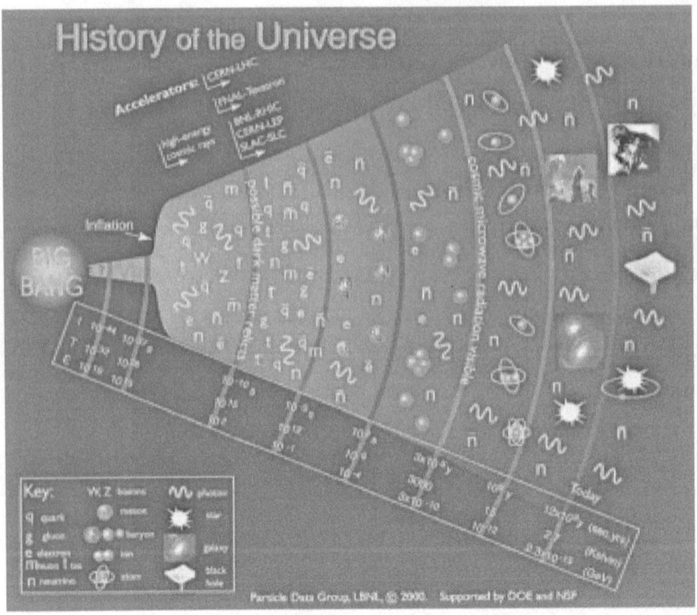

Chapter Seven: Dark Matter, Dark Energy, and the Cosmic Structure

7.1 Introduction: Unveiling the Universe's Hidden Components

ASTROPHYSICAL OBSERVATIONS reveal that the universe's total energy density comprises approximately:

1. **5%** ordinary (baryonic) matter,
2. **27%** dark matter,
3. **68%** dark energy.

While ordinary matter forms stars, planets, and living beings, dark matter and dark energy elude direct detection, inferred only through their gravitational and expansive effects on the cosmos. Understanding these enigmatic components is pivotal for a complete cosmological model.

7.2 Dark Matter: The Invisible Scaffold of the Universe

7.2.1 OBSERVATIONAL Evidence

Dark matter's existence is inferred from phenomena such as:

1. **Galactic Rotation Curves**: Stars in galaxies rotate at speeds suggesting the presence of unseen mass.
2. **Gravitational Lensing**: Light from distant galaxies bends more than visible mass alone can account for.
3. **Cosmic Microwave Background (CMB)**: Fluctuations indicate a significant amount of non-baryonic matter.

7.2.2 Theoretical Models

Various hypotheses aim to explain dark matter:

1. **Weakly Interacting Massive Particles (WIMPs)**: Hypothetical

particles interacting via gravity and possibly the weak force.

2. **Axions**: Light particles arising from solutions to the strong CP problem in quantum chromo dynamics.
3. **Sterile Neutrinos**: Heavier counterparts to known neutrinos, interacting only through gravity.

Despite extensive searches, these particles remain undetected, prompting alternative theories.

7.2.3 Unit Particle Theory (UPT) Perspective

UPT posits that dark matter consists of **unit particles** in a unique rotational state:

1. **Non-Interacting Rotation**: These unit particles lack the rotational configurations necessary for forming standard matter, rendering them invisible.
2. **Gravitational Influence**: Despite their invisibility, their mass contributes to the gravitational potential, influencing galactic structures.
3. **Cosmic Web Formation**: Dark matter unit particles form a scaffold, guiding the distribution of galaxies and large-scale structures.

This framework aligns with observations of dark matter's role in galaxy formation and clustering.

7.3 Dark Energy: The Driver of Cosmic Acceleration

7.3.1 OBSERVATIONAL Evidence

The accelerated expansion of the universe suggests the presence of dark energy, supported by:

1. **Type Ia Supernovae**: These standard candles indicate an increasing rate of cosmic expansion.
2. **Large Scale Structure**: The distribution of galaxies implies repulsive

energy counteracting gravitational attraction.

3. **CMB Measurements**: Data from missions like WMAP and Planck reveal a nearly flat universe, necessitating an additional energy component.

7.3.2 Theoretical Interpretations

Dark energy is often associated with:

1. **Cosmological Constant (Λ)**: A constant energy density filling space homogeneously.
2. **Quintessence**: A dynamic field with evolving energy density.
3. **Modified Gravity Theories**: Alterations to General Relativity at cosmological scales.

7.3.3 UPT Perspective

In UPT, dark energy emerges from the behavior of unit particles:

1. **Rotational Dispersal**: At vast scales, unit particles exhibit an outward rotational motion, generating a repulsive effect.
2. **Energy Density Gradient**: This motion creates a gradient leading to accelerated expansion.
3. **Dynamic Equilibrium**: The interplay between unit particle rotation and density governs the universe's expansion rate.

Recent analyses suggest that dark energy's influence may be evolving over time, indicating a dynamic component rather than a simple cosmological constant.

7.4 Cosmic Structure: A Hierarchy Shaped by Hidden Forces

7.4.1 FORMATION OF the Cosmic Web

The universe's large-scale structure resembles a web, with galaxies residing along filaments and vast voids in between:

1. **Dark Matter's Role:** Acting as a gravitational scaffold, dark matter attracts baryonic matter, leading to galaxy formation along these filaments.
2. **Simulations**: Computational models incorporating dark matter successfully reproduce the observed cosmic web, validating its foundational role.

7.4.2 Influence of Dark Energy

Dark energy affects structure formation by:

1. **Counteracting Gravity**: Its repulsive nature slows the growth of large-scale structures.
2. **Modifying Galaxy Distribution**: Accelerated expansion influences the separation and evolution of galaxy clusters.

7.4.3 UPT Integration

UPT provides a unified view:

1. **Unit Particle Interactions**: The distribution and rotational states of unit particles dictate the formation and evolution of cosmic structures.
2. **Dynamic Interplay**: The balance between attractive forces from dark matter unit particles and repulsive effects from dark energy unit particles shapes the universe's architecture.

7.5 Implications and Future Directions

UNDERSTANDING DARK matter and dark energy is crucial for:

1. **Fundamental Physics**: Insights into these components could lead to new physics beyond the Standard Model.
2. **Cosmology**: Refining models of the universe's past and future evolution.
3. **Astrophysics**: Enhancing knowledge of galaxy formation and

behavior.

Future missions, such as the Euclid Space Telescope, aim to provide high-precision measurements to further constrain the properties of dark matter and dark energy, potentially validating or refuting aspects of UPT.

7.6 Summary

Component	Traditional Understanding	UPT Interpretation
Dark Matter	Composed of unknown particles like WIMPs or axions	Unit particles in non-interacting rotational states forming gravitational scaffolds
Dark Energy	A cosmological constant or dynamic field causing acceleration	Outward rotational dispersal of unit particles creating repulsive energy gradients
Cosmic Structure	Result of gravitational attraction and expansion dynamics	Emergent from the interplay of unit particle densities and rotational behaviors

UPT OFFERS A COHESIVE framework, attributing the universe's hidden mass-energy components and structural formation to the fundamental properties and interactions of unit particles.

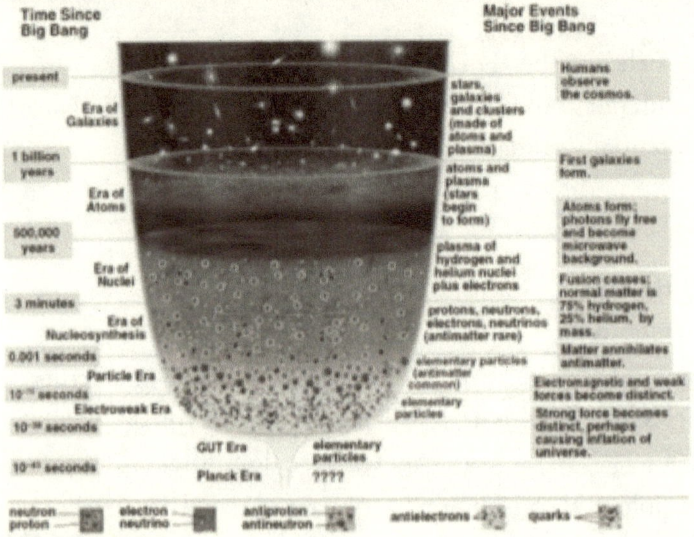
Time Since Big Bang
Major Events Since Big Bang
present
Era of Galaxies
1 billion years
Era of Atoms
500,000 years
Era of Nuclei
3 minutes
Era of Nucleosynthesis
0.001 seconds
Particle Era
Electroweak Era
GUT Era
Planck Era
stars, galaxies and clusters (made of atoms and plasma)
atoms and plasma (stars begin to form)
plasma of hydrogen and helium nuclei plus electrons
protons, neutrons, electrons, neutrinos (antimatter rare)
elementary particles (antimatter common)
elementary particles
elementary particles
????
Humans observe the cosmos.
First galaxies form.
Atoms form; photons fly free and become microwave background.
Fusion ceases; normal matter is 75% hydrogen, 25% helium, by mass.
Matter annihilates antimatter.
Electromagnetic and weak forces become distinct.
Strong force becomes distinct, perhaps causing inflation of universe.
neutron
proton
electron
neutrino
antiproton
antineutron
antielectrons
quarks

Chapter Eight: Unifying All Forces Through Unit Particle Dynamics

8.1 A Century of Separation: The Fragmented View of Forces

FOR OVER A HUNDRED years, physics has classified **four fundamental forces:**

1. **Gravitational** – acting on all mass,
2. **Electromagnetic** – governing charge interactions,
3. **Strong nuclear** – binding atomic nuclei,
4. **Weak nuclear** – causing particle decay.

Each force was modelled with separate **carriers**, **fields**, and **rules**. Despite efforts like the Standard Model and attempts at Grand Unified Theories, the core mystery remains: **Why are these forces separate? Are they truly different—or reflections of the same deeper phenomenon?**

8.2 The UPT Principle: A Single Force Originating from Unit Particles

UNIT PARTICLE THEORY (UPT) reveals a bold conclusion:

All forces are manifestations of the same underlying dynamics of Unit Particles—density, motion, and rotation.

Here's how:

1. **Gravity** emerges from **density gradients** in the rotational fields of Unit Particles.
2. **Electromagnetic forces** are due to **directed flow and field alignment** of rotational energy.
3. **Strong and weak forces** arise from **rotational bonding** and **reconfiguration** at microscopic scales.

Thus, the **variability of force strength and range** is not due to different forces, but due to the **scale and configuration** of rotational interactions between Unit Particles.

8.3 One Law to Bind Them All: Rotational Field Interaction

THE UPT FOUNDATION rests on:

1. **Rotational symmetry** of all matter and fields,
2. **Density-dependent interaction,**
3. **Conservation of angular momentum.**

The strength, reach, and character of a "force" depend on:

1. How tightly Unit Particles are packed (density),
2. Their directional synchronization (rotation axis),
3. Their movement relative to others (momentum).

This explains:

1. **Why gravity is weak but universal**: It operates through **large, diffuse gradients** of Unit Particle density.
2. **Why electromagnetism is stronger**: It involves **polarized and channeled** rotational structures.
3. **Why nuclear forces are the strongest**: They exist in **tightly compact, high-rotation-density zones.**

8.4 Implications for Unified Physics

IF ALL INTERACTIONS stem from the same Unit Particle behaviour, UPT unites:

1. **Classical mechanics** (Newton) and **quantum mechanics** (Schrödinger, Heisenberg),
2. **Relativity** (Einstein) and **quantum field theory** (Feynman),
3. **Cosmology** and **particle physics**, into a **singular materialist model**

governed by:

4. Real, indivisible particles,
5. Rotational dynamics,
6. Measurable and visualisable causes.

This challenges the notion of abstract "forces" and returns physics to **tangible, mechanical explanations.**

8.5 Toward a New Grand Unification

IN UPT:

1. No separate bosons are needed to explain forces.
2. No curved space-time or quantum fluctuations without cause.
3. Every interaction—at every scale—is the result of **real particle behaviour.**

Thus, gravity, light, magnetism, decay, and binding—all emerge from **how Unit Particles arrange, rotate, and transfer momentum.**

This is not merely unification—it is a **reduction** of all complexity to **one elegant structure.**

8.6 Summary: From Many to One

Traditional Force	UPT Interpretation
Gravity	Density gradient of Unit Particle fields
Electromagnetism	Directed flow and rotation of Unit Particles
Strong Nuclear	Interlocked rotation within dense clusters
Weak Nuclear	Rotational reconfiguration and emission

BY EXPLAINING ALL INTERACTIONS as outcomes of **one fundamental cause**, UPT provides the **theory of universal equivalence**—a long-sought key to the complete understanding of the universe.

Chapter Nine: Rethinking Space, Time, and Motion in UPT

9.1 The Traditional View: Spacetime and Relative Motion

IN EINSTEIN'S GENERAL Relativity, **space and time** are interwoven into a four-dimensional **spacetime fabric**. This fabric is warped by mass and energy, giving rise to gravity. Meanwhile, **Special Relativity** introduced the idea that time and space are not absolute but relative, dependent on the observer's motion.

While these theories elegantly explain many phenomena, they rely on **non-material constructs**—spacetime curvature, geodesics, and abstract metrics—rather than tangible physical entities.

9.2 Space in UPT: A Sea of Unit Particles

UPT RADICALLY REDEFINES **space** not as an empty, abstract arena, but as a **field of unbound, freely-rotating Unit Particles**. These particles form the **background material** of the universe.

1. **Void does not exist**; what we perceive as empty space is a dense, low-interaction matrix of Unit Particles.
2. **Spatial geometry** results from **density gradients** and **rotational symmetry** of Unit Particles.
3. The **distribution of these particles** sets the rules for motion, interaction, and field effects.

This implies space is **not empty**, but **structured and active**.

9.3 Time in UPT: The Measure of Rotational Synchronization

IN CONVENTIONAL PHYSICS, time is a parameter used to track change. In UPT, **time** is reinterpreted as:

A measure of **cyclical behaviour** of Unit Particles within a system.

1. A “clock” is a **rotational or oscillatory system** of Unit Particles.
2. Time is **locally defined**, depending on **particle density and energy states.**
3. Time **dilates or contracts** naturally when particles shift configuration or density—this reflects **internal energy rearrangement**, not external curvature of spacetime.

This offers a **material basis for relativity** without abstract spacetime.

9.4 Motion in UPT: Transition of Rotational Patterns

IN NEWTONIAN MECHANICS, motion is movement through space; in relativity, it’s along a geodesic in spacetime. In UPT:

1. **Motion is a transfer of rotational momentum** from one configuration of Unit Particles to another.
2. A particle moves because its **internal spin aligns with a directional rotation pattern** of surrounding Unit Particles.
3. **Inertia** arises from the resistance of ambient Unit Particles to reconfigure.
4. **Acceleration** modifies local rotation density, drawing or repelling neighboring structures.

Thus, all motion is a **rotational transition** within a **connected unit field.**

9.5 Reconciling UPT with Relativity

UPT REPRODUCES KEY relativistic effects without curved spacetime:

1. **Time dilation**: Higher density = slower oscillatory rate = slower “time”.
2. **Length contraction**: Relative motion alters perceived density and rotation axis.
3. **Gravity**: Not curvature, but a **pull into denser Unit Particle fields.**

4. **Light paths bend** not from space warping, but because light's motion aligns with **rotational density gradients.**

This maintains the **empirical success of relativity**, while grounding it in **material particle behaviour.**

9.6 Implications for Physics and Cosmology

THIS REDEFINITION BRINGS powerful implications:

1. **Absolute reference frames** (once discarded) may exist subtly in the structure of Unit Particle fields.
2. **Faster-than-light phenomena** may be reinterpreted as **rotational transfers** within ultra-dense fields.
3. **Cosmic expansion** is driven by repulsive reorganizations, not metric inflation.
4. **Uncertainty** in quantum physics may stem from **incomplete resolution of rotational phases**, not true randomness.

Time, space, and motion become **emergent properties**, not fundamental absolutes.

9.7 Summary: From Abstraction to Materialism

Concept	Traditional Physics	UPT Interpretation
Space	Empty vacuum or curved spacetime	Structured sea of Unit Particles
Time	A dimension or coordinate	Cyclic motion of particles, density-defined
Motion	Movement through space or geodesic path	Reconfiguration of rotational momentum in Unit Particle fields
Relativity	Spacetime-dependent phenomena	Density and spin-dependent behaviours of Unit Particles

UPT REPLACES MATHEMATICAL abstraction with a **tangible, measurable** picture of reality, aligning physical experience with a unified material foundation.

Chapter Ten: Mathematical Framework and Predictive Power of UPT

10.1 Why a New Mathematical Model is Needed

CLASSICAL AND MODERN physics rely on:

1. **Newtonian mathematics** for forces and motion,
2. **Relativistic geometry** for space and time,
3. **Quantum mechanics** with probabilistic wave functions.

However, UPT calls for a **new model grounded in discrete unit entities**, where:

1. All matter and field phenomena arise from **integer-grouped Unit Particles**,
2. Forces emerge from **rotational dynamics** and **density gradients**,
3. Time and motion are outcomes of **spin synchronization and momentum flow**.

A shift from **continuous differential models** to **discrete, integer-based dynamics** becomes necessary.

10.2 Core Parameters of Unit Particles

UPT DEFINES EACH UNIT Particle with:

1. **Fixed mass (m_u)** and **volume (V_u)**,
2. **Intrinsic spin (s_u)**: a quantised angular momentum,
3. **Rotation axis vector ($\vec{R}$)**: orientation in space,
4. **Rotational frequency (f_u)**: defines local time flow.

Unit Particles combine into clusters (x, y, z...) to form composite particles like:

1. **Electron** = x Unit Particles,

2. **Proton** = y Unit Particles ($y > x$),
3. **Neutron** = z Unit Particles.

Where $\mathbf{x, y, z} \in \mathbb{N}$ and stable configurations obey harmonic or symmetric rotation rules.

10.3 Constructing Composite Particles

UPT SUGGESTS A **rule set** for particle formation:

Rotational Conjugation Principle:

Unit Particles must combine such that their spin orientations form **a closed rotational circuit** — a self-sustaining, stable structure.

This implies that:

1. For an **electron**, the x Unit Particles must arrange in a 3D rotational loop,
2. A **proton** requires a higher-order configuration with internal tension balance,
3. **Decay** occurs when the loop breaks symmetry, ejecting excess Unit Particles.
4. These rules replace quantum fields with **geometric rotational logic**.

10.4 Fields and Forces as Mathematical Rotational Gradients

FIELDS ARE NOT VECTOR abstractions but **real particle arrangements**:

1. A **gravitational field** is a **density slope** of Unit Particles,
2. An **electric field** is a **spin-aligned layer** of polarized Unit Particles,
3. A **magnetic field** is a **toroidal current loop** of rotating configurations.

Mathematically:

1. Let $\rho_u(\mathbf{r})$ be the Unit Particle density at position **r**,

2. Let $\mathbf{\Omega_u}(\mathbf{r})$ be the rotational vector field of local Unit Particle spin,

Then force **F** on a composite structure is:

$$\mathbf{F} = -\nabla(\rho_u \cdot \Omega_u)$$

A gradient of rotational density multiplied by spin alignment — a **material analogue** to vector field calculus.

10.5 Predictive Capacity of UPT

UPT ALLOWS CALCULATION of:

1. **Mass ratios**: proton-to-electron mass from (y/x),
2. **Charge effects**: from the **directionality of rotational spin flow**,
3. **Field strength**: by density gradients and angular divergence of Ω_u,
4. **Particle decay**: when local symmetry thresholds break,
5. **Gravitational acceleration**: as a result of **unit density pull**, not geometric curvature.

This model supports quantised values **without randomness**, suggesting:

1. Predictability of particle properties from configuration,
2. Derivable constants from base Unit Particle values.

10.6 Limitations and Opportunities

CHALLENGES:

1. Formulating a complete Lagrangian or Hamiltonian structure for UPT,
2. Bridging UPT with quantum statistics (Pauli exclusion, entanglement),
3. Mapping classical field equations onto discrete rotational systems.

Opportunities:

1. Simulating matter formation in digital lattice models,
2. Connecting cosmological constants to Unit Particle field curvature,
3. Deriving Standard Model properties from first principles.

10.7 SUMMARY

Concept	Classical Physics	UPT Formulation
Matter	Continuous field excitations	Integer-grouped Unit Particles
Fields	Mathematical vectors	Rotational density gradients
Forces	Geometric or field interactions	Momentum from spin and density slopes
Time	Parametric coordinate	Rotational frequency of Unit Particles
Motion	Path through space	Reorientation of spin flow
Equations	Differential calculus	Integer-based rotational dynamics

---

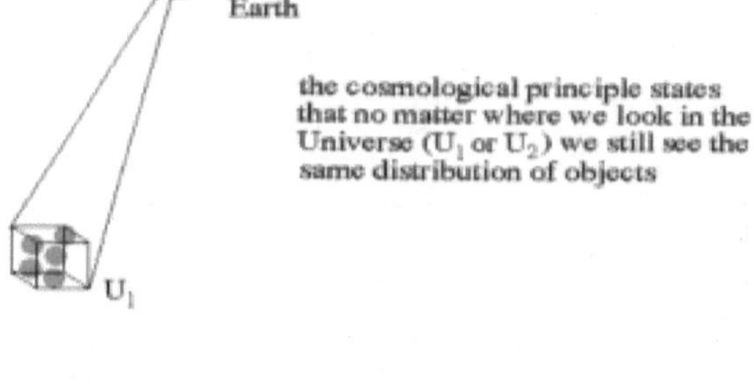

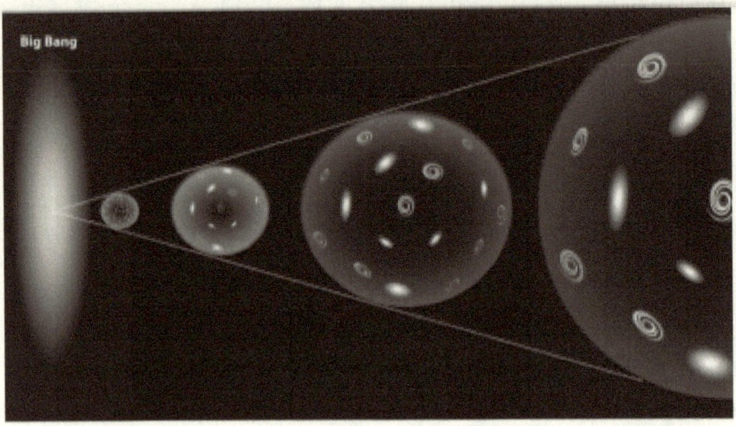
Big Bang

Chapter Eleven: Implications for Quantum Mechanics and Particle Physics

11.1 The Quantum Landscape: Challenges and Interpretations

QUANTUM MECHANICS GOVERNS the behaviour of matter at microscopic scales, relying on concepts like:

1. **Wave-particle duality**
2. **Uncertainty Principle**
3. **Quantum entanglement**
4. **Probability-based predictions**

Yet, despite its experimental success, quantum theory:

1. Offers **no physical model** for wave functions,
2. Demands **non-locality** in entanglement,
3. Cannot easily unify with gravity or General Relativity.

UPT offers a **materialist reinterpretation** of these phenomena using the tangible behaviour of **Unit Particles.**

11.2 Quantum States as Rotational Configurations

IN UPT, A **quantum state** is not a superposition of possibilities but a **real rotational configuration** of Unit Particles.

1. **Superposition**: A system exists in multiple rotational phase states until external interaction locks it into a stable arrangement.
2. **Wave functions**: Mathematical descriptions of **rotational momentum probabilities** in 3D space.
3. **Collapse**: Occurs when a composite system transfers rotational momentum externally—returning to a lower-energy stable configuration.

Thus, randomness in quantum behaviour is **epistemic**, not fundamental—uncertainty arises from incomplete knowledge of **rotational alignments.**

11.3 Spin, Identity, and the Pauli Principle

IN QUANTUM MECHANICS:

1. Particles have **intrinsic spin**,
2. Fermions obey the **Pauli Exclusion Principle**, while bosons do not.

In UPT:

1. **Spin** is the literal angular momentum of Unit Particles,
2. The **Pauli Principle** emerges naturally: two identical composite particles cannot occupy the same rotational phase state in a shared field without inducing **rotational interference**—leading to instability.

Thus, exclusion is not a rule—it is a **natural property of rotational harmony**.

11.4 Entanglement: Rotational Synchronization Across Fields

QUANTUM ENTANGLEMENT baffles with its **non-local correlations**. In UPT:

Entangled particles are **rotationally synchronized substructures** of a shared Unit Particle cluster.

Even when spatially separated, their **spin states remain coupled** through the **rotational field continuum**—not through empty space but via **connected Unit Particle density**, allowing instantaneous alignment without information travel.

No paradox—just **field continuity** in a structured material substrate.

11.5 Quantum Tunnelling and Energy Transitions

QUANTUM TUNNELING DESCRIBES particles crossing energy barriers they classically shouldn't. In UPT:

1. Barriers are **dense Unit Particle fields**,
2. Tunnelling occurs when **rotational momentum temporarily synchronizes** across a density discontinuity, allowing momentary passage.

Likewise, energy level transitions in atoms occur when **spin configurations shift harmonics**, releasing or absorbing rotation packets (interpreted as photons).

11.6 Particle Families and the Standard Model

UPT REINTERPRETS PARTICLE families:

1. **Leptons and quarks** are **meta-structures** of Unit Particle arrangements,
2. Particle "flavors" represent different **rotational symmetry classes**,
3. Mass and charge derive from **Unit count** and **net spin alignment**.

This offers a path toward:

1. **Explaining why particles exist** in discrete types,
2. Deriving Standard Model constants from **first principles**,
3. Unifying interactions through **spin-density evolution**, not separate forces.

11.7 Summary Table: Quantum vs. UPT Views

Quantum Concept	Quantum Mechanics	UPT Interpretation
Wavefunction	Probability amplitude	Rotational phase distribution of Unit Particles
Spin	Intrinsic, abstract property	Real angular momentum of structured Unit Particles
Uncertainty	Fundamental limitation	Result of unknown rotational state
Entanglement	Non-local correlation	Persistent synchronization in a shared spin field
Particle Identity	Defined by quantum numbers	Emergent from integer-based rotational architecture

Chapter Twelve: Cosmological Consequences and the Fate of the Universe

12.1 Cosmology Through the Lens of Unit Particle Theory

CONVENTIONAL COSMOLOGY—BASED on General Relativity—describes the universe in terms of spacetime geometry, expansion, dark energy, and inflation. However, UPT offers a radically different approach grounded in:

1. The **material nature of fields**,
2. The **density and configuration of Unit Particles**,
3. The **rotational logic of structure and motion.**

In this view, the cosmos is not shaped by the bending of abstract space but by the **real distribution and dynamics of Unit Particles.**

12.2 The Birth of the Universe: A Rotational Genesis

UPT POSITS A UNIVERSE originating not from a singularity but from a **highly concentrated, rotationally symmetric field** of Unit Particles.

1. The **initial state** was a dense, uniform rotational lattice.
2. A **symmetry disruption**—not an explosion—caused spinning regions to separate and diversify, leading to matter formation.
3. Expansion is the **spatial unfolding of rotating density zones,** not the stretching of space itself.

This model sidesteps the paradox of infinite density and offers a **real, physical origin.**

12.3 Galactic Formation and Structure

GALAXIES, STARS, AND planetary systems form as **self-organising rotational systems** of Unit Particles:

1. **Spiral arms** reflect density waves of angular momentum,
2. **Central bulges** are high-density vortex cores,
3. Gravitational pull is a **net flow toward rotational centers**, driven by Unit Particle density gradients.

This provides a physical mechanism for:

1. Galaxy rotation curves,
2. Disk formation,
3. Stability across vast timescales—without needing exotic matter.

12.4 A Material Explanation for Dark Matter

INSTEAD OF POSTULATING invisible particles, UPT suggests:

Dark matter is composed of Unit Particle configurations that do not emit light—due to their internal spin symmetry, they do not interact electromagnetically.

They:

1. Contribute **mass and gravity** via field density,
2. Form **halos** around galaxies,
3. Are **detectable only by gravitational influence**, not radiation.

This demystifies dark matter as simply **undetectable but real material structures.**

12.5 Understanding Dark Energy and Expansion

UPT REINTERPRETS COSMIC acceleration not as "repulsive energy" but as:

1. A **field-level redistribution** of rotational momentum,
2. Outer layers of the Unit Particle field continuing to **unfold** and expand,
3. **Apparent acceleration** caused by decreasing central density and

inertia resistance.

No new force is needed—just the consequences of **field evolution and spin tension balance.**

12.6 The Fate of the Universe: Rotational Equilibrium or Collapse

RATHER THAN HEAT DEATH or eternal inflation, UPT proposes:

1. A final state of **rotational equilibrium**, where all Unit Particle fields stabilise into **non-violent symmetry**,
2. Or a **re-coalescence**, where declining outer densities reverse into inward spin flows—**a cosmic rebirth** through rotational contraction.

Cycles may arise from:

1. **Spin reorganisation thresholds,**
2. **Global resonance resets,**
3. **Density synchronization events.**

12.7 Summary and Outlook

Cosmological Concept	Standard Model View	UPT Perspective
Origin	Big Bang singularity	Rotational symmetry breaking in a dense field
Structure formation	Matter clumps via gravity	Vortex formation in rotational field density
Dark matter	Unknown particles	Non-radiating Unit Particle clusters
Dark energy	Repulsive energy, cosmological constant	Rotational tension and momentum dispersion
Expansion	Stretching of space	Unfolding of rotating field layers
Fate of Universe	Heat death, Big Rip	Rotational equilibrium or cyclic rebirth

---

CONCEPTUAL DIAGRAM: Rotational Genesis and Cosmic Evolution in UPT

1. **Initial Uniform Rotational Field:**
   - Depict a dense, uniform field of Unit Particles, each represented as small spheres with synchronized rotational arrows, symbolizing the initial symmetrical state of the universe.
2. **Symmetry Disruption and Formation of Vortices:**
   - Illustrate a disturbance causing certain regions to develop varied rotational orientations, leading to the creation of vortices. These vortices represent the nascent formations of matter and energy concentrations.
3. **Development of Galactic Structures:**
   - Show how these vortices evolve into larger rotational systems, forming spiral galaxies. Arrows can indicate the

direction of rotational momentum, emphasizing the self-organizing nature of these structures.

4. **Dark Matter as Non-Radiating Unit Particle Clusters**:
    - Include halo-like structures around galaxies composed of Unit Particles without electromagnetic interactions, representing dark matter. These should be distinguishable yet integrated into the overall rotational framework.
5. **Cosmic Expansion Through Rotational Unfolding**:
    - Illustrate the outward unfolding of rotational layers, symbolizing the expansion of the universe driven by rotational momentum dispersion rather than spatial stretching.

This conceptual diagram would encapsulate the core principles of UPT, providing a visual representation of the universe's evolution through rotational dynamics of Unit Particles.

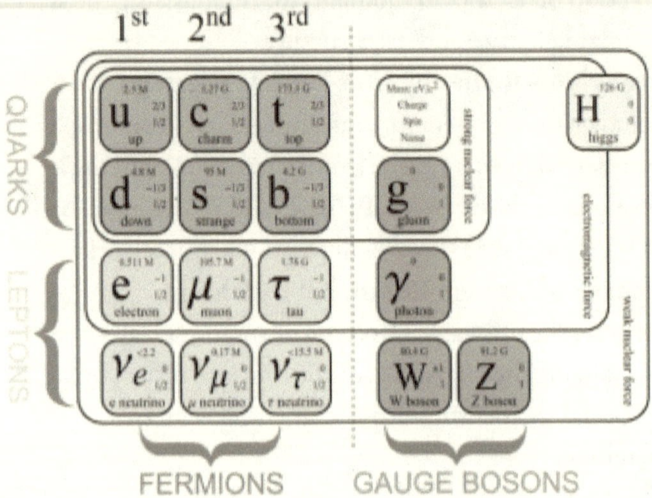
1st 2nd 3rd
QUARKS
LEPTONS
u up
c charm
t top
d down
s strange
b bottom
e electron
μ muon
τ tau
νe e neutrino
νμ μ neutrino
ντ τ neutrino
Mass Charge Spin Name
g gluon
γ photon
W W boson
Z Z boson
H higgs
strong nuclear force
electromagnetic force
weak nuclear force
FERMIONS
GAUGE BOSONS

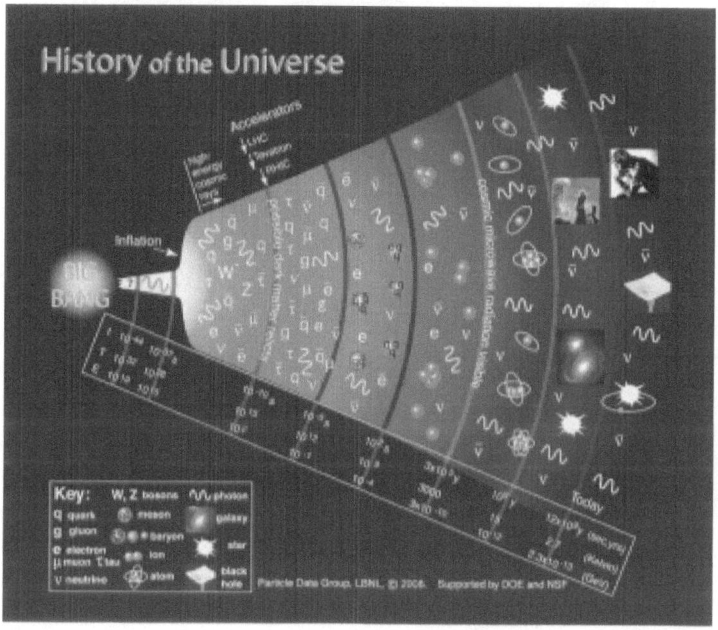
History of the Universe
Accelerators
LHC
Tevatron
RHIC
Inflation
BIG BANG
Today
Key:
W, Z bosons
photon
quark
meson
galaxy
gluon
baryon
electron
star
muon
tau
ion
neutrino
atom
black hole
Supported by DOE and NSF

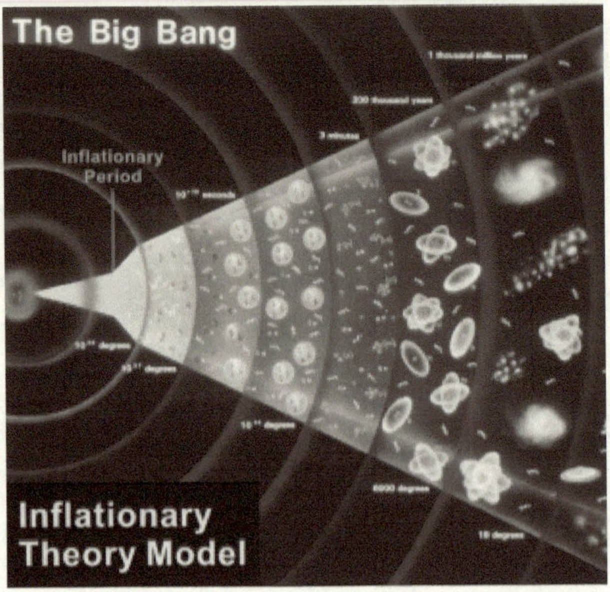
The Big Bang
Inflationary
Period
Inflationary
Theory Model

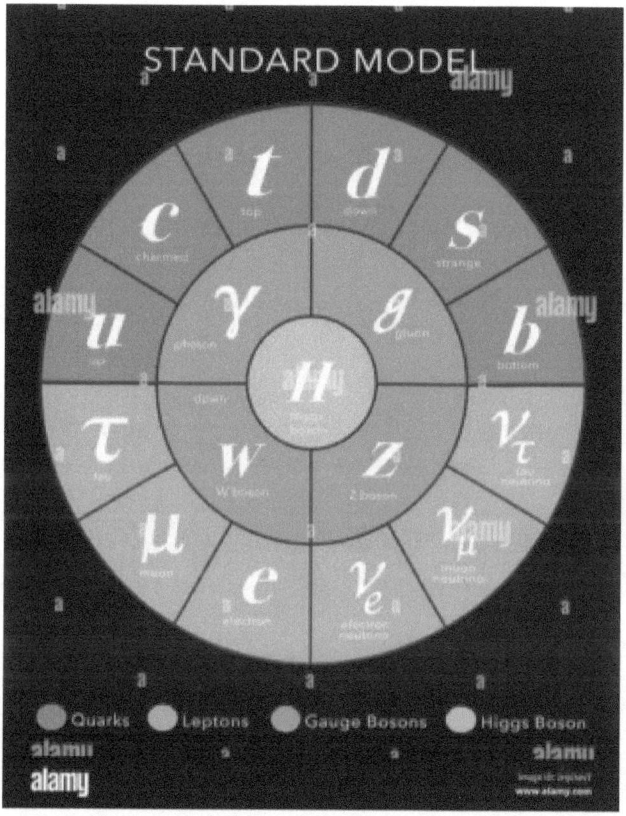
STANDARD MODEL
t
d
c
s
u
b
γ
g
H
τ
W
Z
μ
e
Quarks
Leptons
Gauge Bosons
Higgs Boson

COSMOLOGICAL HORIZON
UNIVERSE
Earth
Speed of Light

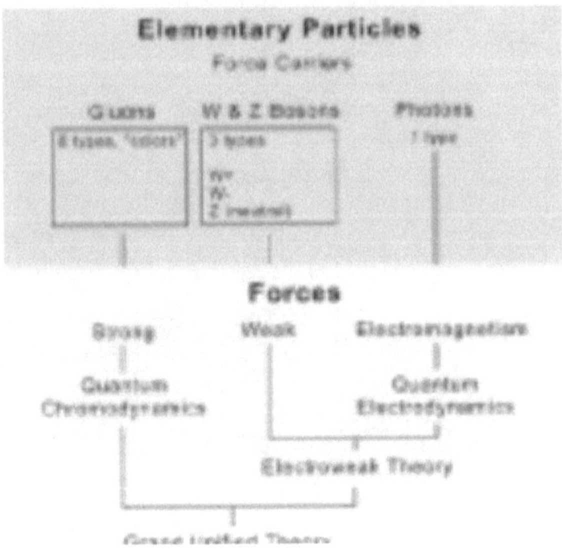
Elementary Particles
Force Carriers
Gluons
8 types, "colors"
W & Z Bosons
3 types
W+
W-
Z (neutral)
Photons
1 type
Forces
Strong
Weak
Electromagnetism
Quantum Chromodynamics
Quantum Electrodynamics
Electroweak Theory

Chapter Thirteen: Unifying Forces Under Unit Particle Theory

13.1 Revisiting the Fundamental Forces

CONTEMPORARY PHYSICS recognises four fundamental forces:

1. **Gravitational force**
2. **Electromagnetic force**
3. **Strong nuclear force**
4. **Weak nuclear force**

These forces are treated as separate interactions mediated by unique particles—gravitons (hypothetical), photons, gluons, and W/Z bosons, respectively. The Standard Model successfully describes three of these but fails to include gravity. Moreover, a deeper connection between them remains elusive.

Unit Particle Theory (UPT), however, proposes a new paradigm: **all forces arise from variations in the rotational behaviour and density of Unit Particles.**

13.2 The Common Origin: Rotational Fields and Density Gradients

ACCORDING TO UPT, EVERY observable force is a **manifestation of rotational momentum and configuration among Unit Particles**. Fields are not abstract mathematical constructs, but real, physical distributions of rotating Unit Particles. Here's how each force maps to this concept:

1. **Gravitational Force**: Emerges from **density gradients** in rotational Unit Particle fields; bodies are drawn toward regions of higher Unit Particle density.
2. **Electromagnetic Force**: Arises from **alignment and oscillation** of rotational moments across charged Unit Particle systems.

3. **Strong Force:** Occurs when **adjacent Unit Particle clusters (like quarks)** achieve a dense rotational lock-in that resists separation.
4. **Weak Force:** Is an expression of **rotational phase transitions** between composite particle structures—explaining phenomena like beta decay.

13.3 Equivalence in Action

BY EXAMINING HOW DIFFERENT forces act, UPT reveals their shared traits:

1. All forces have **finite ranges** depending on rotational coherence.
2. All involve **momentum transfer**, not the transmission of abstract "charges."
3. Apparent differences arise from **scale and complexity of Unit Particle arrangements.**

This removes the artificial division of force types. Rather than four disparate forces, nature has **one unified dynamic**: the behaviour of structured rotational fields.

13.4 Rethinking Force-Carrying Particles

STANDARD MODEL FORCES are "mediated" by particles. UPT reinterprets these:

1. **Photons**: Packets of rotational momentum transferred between charged Unit Particle structures.
2. **Gluons**: Short-range connectors within rotational clusters, reinforcing strong-field cohesion.
3. **W/Z Bosons**: Products of temporary spin symmetry breaks—field phase reconfigurations—not fundamental particles.

No separate mediators are required—**Unit Particles themselves carry and respond to all interactions.**

13.5 Gravity Without Curved Space

A KEY TRIUMPH OF UPT is its explanation of gravity **without invoking spacetime curvature**:

Gravity is not a geometrical effect, but a physical consequence of **spin density differences in Unit Particle fields**.

The pull of gravity becomes a tendency of systems to move toward regions of **higher rotational density**, much like pressure gradients in a fluid.

This brings gravity into the same framework as other forces—at last, a true **unification**.

13.6 Implications for the Grand Unified Theory (GUT)

UPT POTENTIALLY SUPERSEDES traditional GUT attempts by:

1. Providing a **mechanical foundation** for all interactions,
2. Eliminating the need for energy-based unification points,
3. Enabling the derivation of force laws from **Unit Particle architecture**.

It opens the door to:

1. Deriving constants (e.g. gravitational constant, Planck's constant) from Unit Particle parameters,
2. Explaining why interactions differ across scales,
3. Designing new experiments based on spin field manipulation.

13.7 Summary Table: Unified Force Interpretation

Force Type	Standard View	UPT Interpretation
Gravity	Curved spacetime	Spin density gradient in Unit Particle field
Electromagnetism	Field mediated by photons	Spin oscillations and field tension
Strong Force	Gluon-mediated quark binding	Dense spin interlock in Unit Particle clusters
Weak Force	Boson-mediated particle decay	Rotational phase transition in composite systems
Carrier Particles	Abstract mediators	Emergent states of Unit Particle momentum

CONCEPTUAL DIAGRAM: Force Unification via Rotational Fields in UPT

1. **Unit Particles as Fundamental Entities**:
    - Represent Unit Particles as small spheres, each with an arrow indicating its intrinsic spin or rotational momentum.
2. **Formation of Rotational Fields**:
    - Illustrate groups of Unit Particles aligning their spins to form larger rotational fields. These fields represent the emergence of different fundamental interactions based on their configurations.
3. **Depiction of Fundamental Forces**:
    - **Gravitational Force**: Show a large-scale, coherent rotational field where the collective spin alignment of Unit Particles creates a macroscopic attraction, symbolizing gravity.
    - **Electromagnetic Force**: Illustrate smaller, more localized rotational fields with opposing spin orientations, representing dipole formations that give rise to electromagnetic interactions.
    - **Strong Nuclear Force**: Depict tightly packed Unit Particles with interlocking spins, indicating the short-range but

powerful attraction holding atomic nuclei together.
    - **Weak Nuclear Force**: Show a scenario where a Unit Particle undergoes a spin reorientation, leading to a change in particle identity, symbolizing processes like beta decay.
4. **Unification through Rotational Symmetry**:
    - Encompass all these interactions within a larger framework that highlights the underlying rotational symmetry of Unit Particles, emphasizing that all fundamental forces originate from variations in spin configurations and alignments.

This conceptual diagram would encapsulate the essence of UPT's approach to unifying fundamental forces through the dynamics of rotational fields formed by Unit Particles.

Force Unification Theories, the Standard Model, and the Early Universe

These Feynman diagrams depict weak interactions.

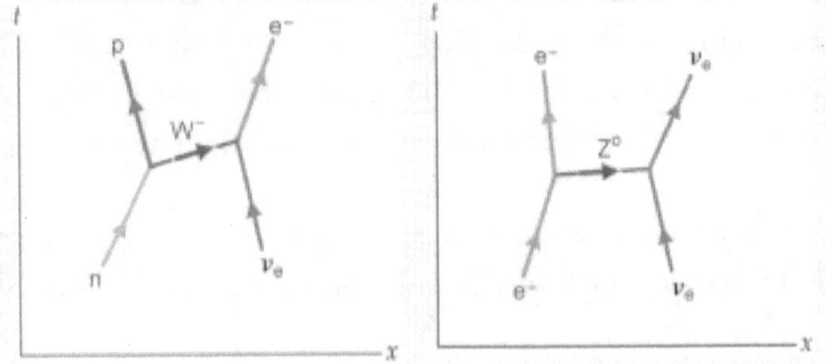

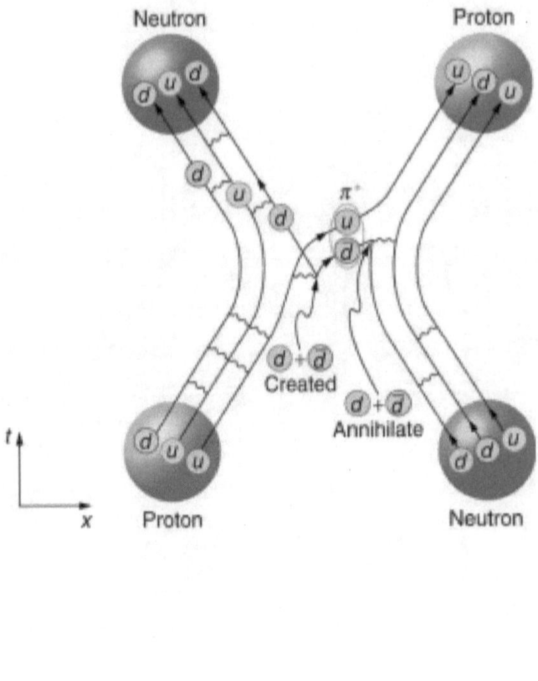

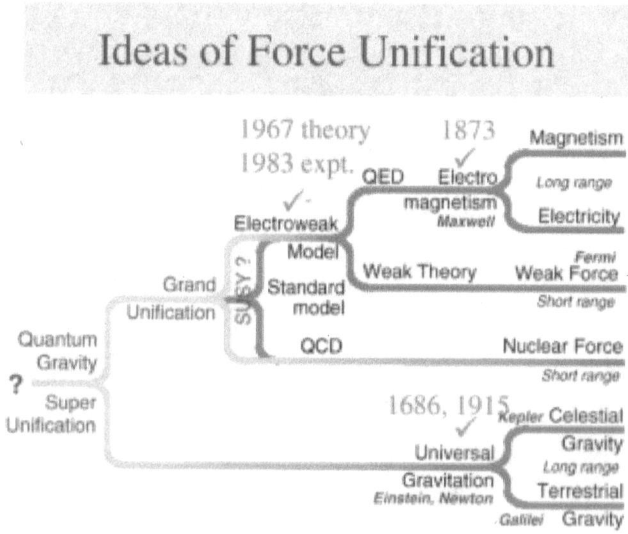

Grand Unification Scheme

- To explain forces in terms of exchanging particles

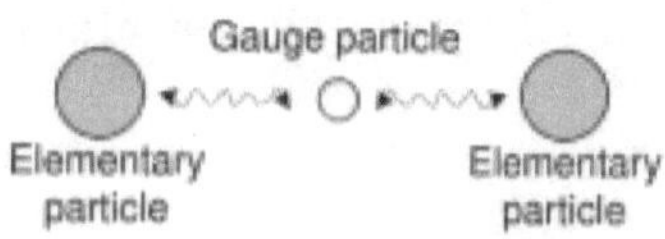

Each coupling constant (related to a specific force) converges to one value.

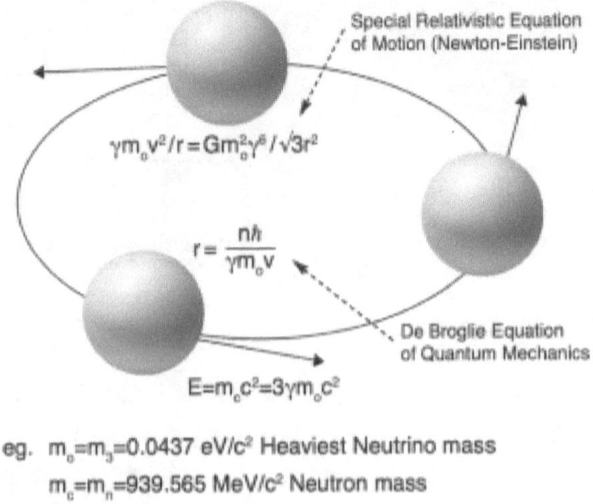
Special Relativistic Equation
of Motion (Newton-Einstein)
$\gamma m_o v^2/r = Gm_o^2\gamma^6/\sqrt{3}r^2$
$r = \frac{n\hbar}{\gamma m_o v}$
De Broglie Equation
of Quantum Mechanics
$E=m_c c^2=3\gamma m_o c^2$
eg. $m_o=m_3=0.0437$ eV/c^2 Heaviest Neutrino mass
$m_c=m_n=939.565$ MeV/c^2 Neutron mass

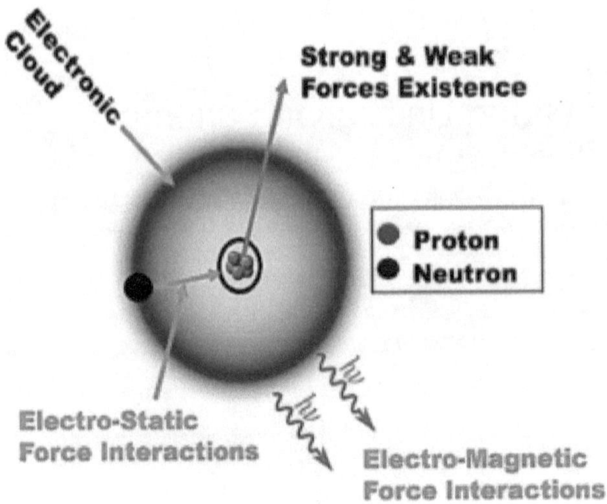
Electronic
Cloud
Strong & Weak
Forces Existence
Proton
Neutron
Electro-Static
Force Interactions
Electro-Magnetic
Force Interactions

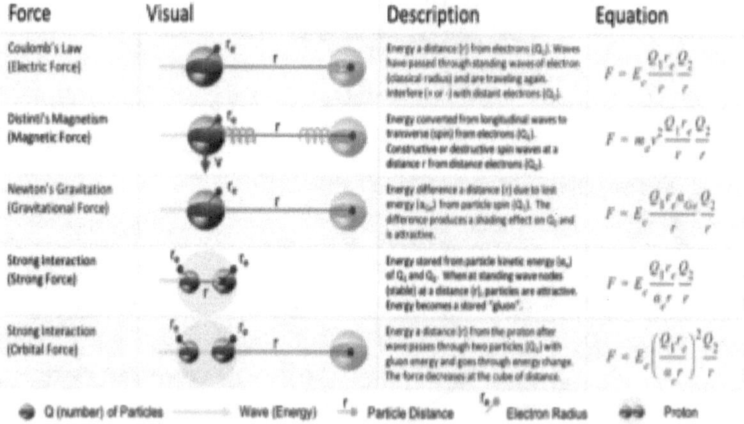
Force
Visual
Description
Equation
Coulomb's Law (Electric Force)
Energy a distance (r) from electrons (Q1). Waves have passed through standing waves of electron (classical radius) and are traveling again. Interfere (+ or -) with distant electrons (Q2).
Distinti's Magnetism (Magnetic Force)
Energy converted from longitudinal waves to transverse (spin) from electrons (Q1). Constructive or destructive spin waves at a distance r from distance electrons (Q2).
Newton's Gravitation (Gravitational Force)
Energy difference a distance (r) due to lost energy from particle spin (Q1). The difference produces a shading effect on Q2 and is attractive.
Strong Interaction (Strong Force)
Energy stored from particle kinetic energy of Q1 and Q2. When at standing wave nodes (stable) at a distance (r), particles are attraction. Energy becomes a stored "gluon".
Strong Interaction (Orbital Force)
Energy a distance (r) from the proton after wave passes through two particles (Q1) with gluon energy and goes through energy change. The force decreases at the cube of distance.
Q (number) of Particles
Wave (Energy)
Particle Distance
Electron Radius
Proton

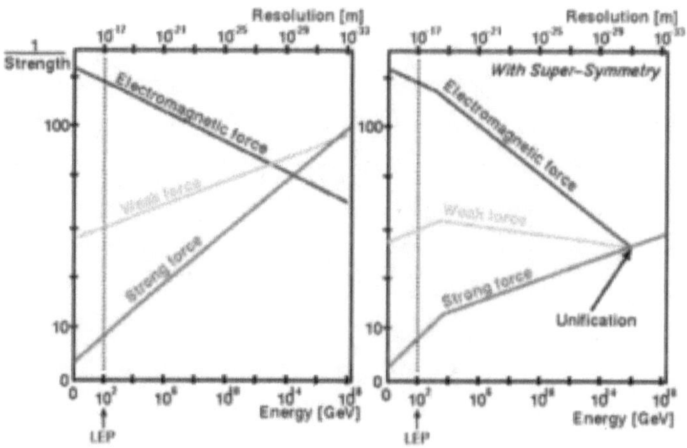
Resolution [m]
1 / Strength
Electromagnetic force
Weak force
Strong force
Energy [GeV]
LEP
With Super-Symmetry
Unification

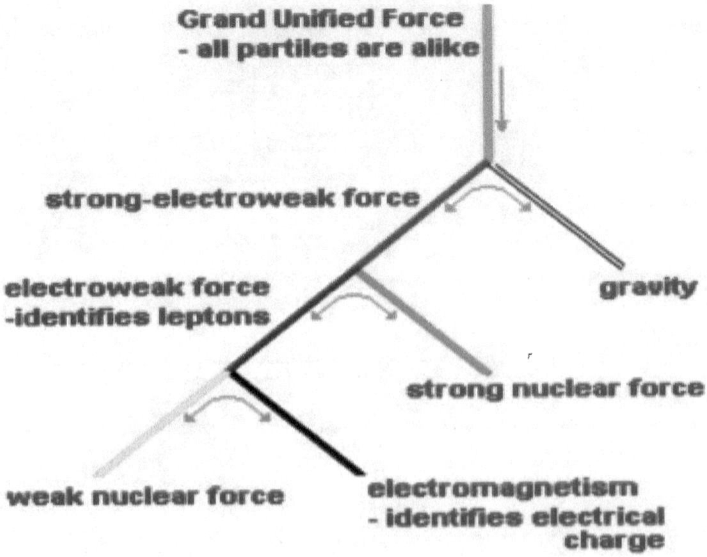
Grand Unified Force
- all partiles are alike
strong-electroweak force
gravity
electroweak force
-identifies leptons
strong nuclear force
weak nuclear force
electromagnetism
- identifies electrical charge

Chapter Fourteen: Technological and Experimental Implications of Unit Particle Theory

14.1 Introduction to Practical Potential

UNIT PARTICLE THEORY (UPT), by offering a fundamental model of matter and force interactions grounded in the behaviour of indivisible Unit Particles, opens a revolutionary path in applied science. This chapter explores the **technological, experimental, and societal ramifications** of UPT—how it may inform future breakthroughs, redefine physical experiments, and enable new capabilities in energy, communication, and materials science.

14.2 Experimental Validation of UPT

TO TRANSITION UPT FROM theory to recognized scientific framework, it must be subjected to falsifiable, measurable experimentation. Several promising avenues include:

A. High-Resolution Density Mapping

1. Experiments using interferometer or atomic-resolution imaging could test predicted **variations in Unit Particle field density** surrounding masses.
2. If gravitational pull correlates with **rotational density** rather than simply mass, this would favour UPT.

B. Spin Resonance in Synthetic Particles

1. UPT predicts that **specific configurations of spin-aligned Unit Particles** generate fundamental particles (e.g., electrons, protons).
2. By manipulating ultra-cold atomic lattices, we could attempt to **synthesize particles from pre-arranged spin clusters**, testing UPT's composition model.

C. Revisiting Cavendish-Style Experiments

1. Redoing gravity experiments using highly sensitive torsion balances in different **rotational environments or temperature states** may uncover gravitational variations consistent with Unit Particle field dynamics.

14.3 UPT-Inspired Technologies

UPT IMPLIES THAT CONTROLLING the orientation and density of Unit Particles could **unlock new forms of energy and matter manipulation**.

A. Energy Engineering

1. Harnessing rotational momentum in dense Unit Particle fields may offer a new class of **rotational energy devices**, potentially more efficient and scalable than traditional mechanics or even nuclear energy.

B. Propulsion Systems

1. If gravity is a function of density gradients in spin fields, UPT suggests the possibility of **mass manipulation or inertial control**.
2. Spacecraft might eventually use controlled spin-field emitters to "lean" into gravitational paths.

C. Information Transfer and Field Computing

1. Unit Particle fields may allow **field-based data encoding**, where information is transferred not through electromagnetic waves, but via modulated **spin densities**.
2. Such technology could surpass quantum computing in speed and complexity.

14.4 Redefining the Scientific Paradigm

A. Elimination of Force Carriers

1. Current models rely on mediators like photons or gluons. UPT proposes a direct-field interaction model.
2. If proven, this could simplify particle physics, cutting layers of abstraction and **offering a clearer, more deterministic universe.**

B. A Mechanical Basis for Reality

1. UPT offers a return to classical-style thinking—not of hard particles, but of **continuous, causally structured fields** made of real entities.
2. This contrasts with probabilistic quantum mechanics, instead promoting a **deterministic foundation for physical law.**

14.5 Societal Impacts and Philosophical Shifts

A. Educational Transformation

1. Physics education could be rewritten to **begin with field mechanics** and **particle emergence**, rather than treating particles as irreducible units.

B. Ethical Implications

1. As with any paradigm-shifting theory, UPT's realization could raise ethical concerns—especially with regard to **field manipulation technologies** capable of altering matter or inertia.

C. Philosophical Significance

1. UPT suggests an **underlying unity** not just of matter, but of force, space, and motion—echoing ancient materialist and metaphysical doctrines in a scientifically rigorous way.

14.6 Future Directions

1. **Research Institutes**: Dedicated UPT centers can investigate rotational field dynamics, simulate spin field interactions, and prototype new instruments.
2. **Global Collaborations**: Just as CERN unified physicists for the Standard Model, a UPT-focused consortium could unite researchers across disciplines—physics, engineering, materials science, and cosmology.

14.7 Conclusion

THE EXPERIMENTAL AND technological implications of Unit Particle Theory are vast. It not only proposes a simpler, unified view of forces and particles, but also lays the groundwork for **real-world applications** that could redefine physics, energy, and information. As research progresses, UPT may transition from a bold theory to the **operational framework of a new scientific age**.

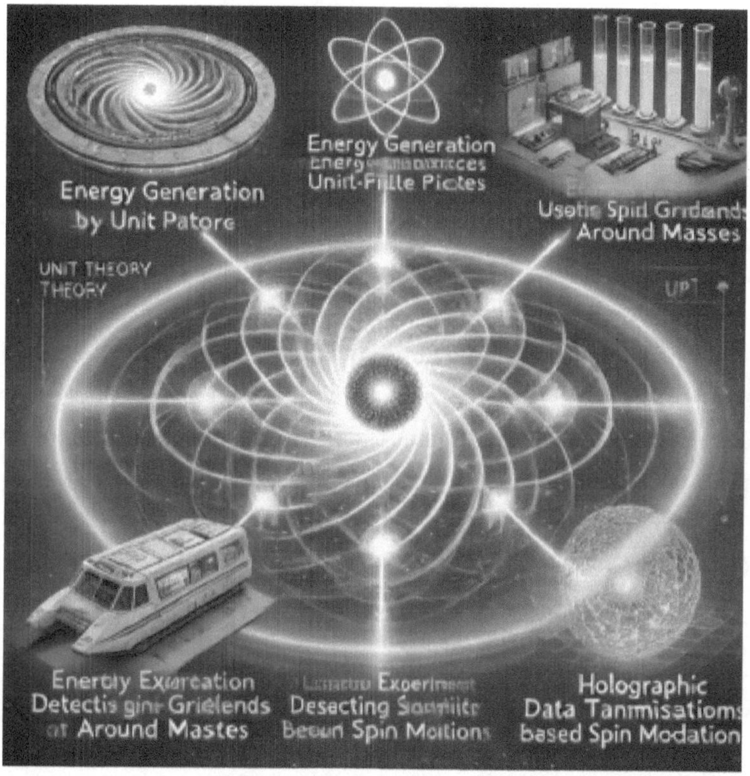
Energy Generation
by Unit Patore
Energy Generation
Unirl-Fille Pictes
Usetis Spid Gridands
Around Masses
UNIT THEORY
THEORY
Enertly Exureation
Detectis gin Grielends
at Around Mastes
Experiment
Desecting Sonnilte
Beoun Spin Motions
Holographic
Data Tanmisatioms
based Spin Modation

Chapter Fifteen: Cosmological Model Based on Unit Particle Theory

15.1 Introduction

UNIT PARTICLE THEORY (UPT) provides not only a microscopic explanation of particles and forces, but also offers a **macroscopic vision of the universe itself.** By understanding cosmic structures as large-scale configurations of Unit Particles and their spin-based fields, UPT proposes a **novel cosmological model**—one in which galaxies, stars, and even the expansion of the universe are direct consequences of underlying field dynamics and rotational symmetry.

15.2 A Universe Built from Rotational Fields

IN UPT, ALL MATTER is composed of **rotating Unit Particles** grouped into structures by integer multiples. These micro-rotational fields extend naturally into the cosmic scale:

1. **Galaxies** form as **large rotational formations**, where vast numbers of Unit Particles are aligned in dense field configurations.
2. **Stars** arise in localized field collapses where rotational momentum achieves nuclear densities.
3. **Dark matter and dark energy** are reinterpreted as **low-density Unit Particle fields**—invisible but structurally influential.

15.3 Gravity as a Gradient of Field Density

TRADITIONAL MODELS describe gravity as the curvature of spacetime, but UPT asserts:

1. **Gravitational attraction** results from the **gradient in Unit Particle field density**, especially due to rotational compression around mass centers.
2. Massive cosmic structures attract others because they **compress**

surrounding Unit Particle fields, forming density wells.

3. This offers an **alternative explanation for gravitational lensing and galactic orbit speeds**, currently attributed to dark matter.

15.4 The Expansion of the Universe

UPT PROVIDES A MECHANICAL explanation for the observed cosmological expansion:

1. Initially, all Unit Particles existed in a **high-density rotational matrix**, balanced in perfect symmetry.
2. A disturbance—perhaps quantum fluctuation or a local alignment imbalance—disrupted the equilibrium, leading to **field fragmentation.**
3. These broken fields **rotated outward**, initiating large-scale **repulsive interactions** due to inverse density gradients.

Thus, cosmic expansion arises not from a spacetime inflation, but from the **rotational reconfiguration of Unit Particle fields** at a universal scale.

15.5 Cosmic Background Radiation and Field Resonance

THE **Cosmic Microwave Background (CMB)** is traditionally viewed as a remnant of the Big Bang. In UPT:

1. The CMB may instead be interpreted as the **resonant noise** of Unit Particle spin-field interactions in the vacuum.
2. This radiation is stable and uniform because it emerges from a **fundamental background field**—the lowest-energy state of Unit Particle motion across the cosmos.

This interpretation removes the need for inflation theory and supports a **steady but evolving field-based universe.**

15.6 Time and Space as Emergent Quantities

IN UPT:

1. **Time** is not absolute but a **function of field rotational frequency**—the higher the local spin density, the slower the passage of time.
2. **Space** is defined not as an empty container, but as the **dynamic scaffold formed by interconnected rotational fields** of Unit Particles.

This creates a **relational universe**, where time and distance are properties of field behaviour rather than fixed coordinates.

15.7 Implications for Cosmology

A. Galaxy Formation

1. Spiral galaxies naturally arise from **coherent rotational fields**, explaining their stable form without invoking invisible matter halos.

B. Cosmic Structure

1. The large-scale web of the universe mirrors the **interlinked Unit Particle fields**, where voids are low-density zones and filaments represent high-density spin currents.

C. Redshift and Energy Loss

1. Light loses energy not by stretching through expanding space but via **interaction with the rotating Unit Particle field**, creating a Redshift analogous to resistance in a medium.

15.8 Conclusion

UPT REIMAGINED THE universe not as a chaotic explosion from a singularity, but as an elegant, self-rotating **continuum of structured fields.** It integrates **particle physics, force interaction, and cosmology** under one universal mechanism: the rotation and arrangement of indivisible Unit Particles. This model not only provides a deeper understanding of observed phenomena but opens the path for future refinement of cosmological theories through a **mechanically grounded, field-based vision of the universe.**

Chapter Sixteen: Predictive Power and Theoretical Tests of Unit Particle Theory

16.1 Introduction

A THEORY'S STRENGTH lies in its ability to **predict observable phenomena** and suggest **testable hypotheses**. Unit Particle Theory (UPT), while offering a unified model of matter and force, also provides clear predictions that distinguish it from the Standard Model and General Relativity. This chapter outlines those predictions and suggests experimental setups or natural observations that could validate—or challenge—UPT.

16.2 Quantised Composition of Particles

A. Integer-Based Particle Construction

UPT POSTULATES THAT all known particles are composed of **whole-number groupings of Unit Particles**. For example:

1. An **electron** is formed by x Unit Particles.
2. A **proton** is formed by y Unit Particles, where $y > x$.

This yields several **testable predictions**:

1. **Exact mass ratios** between particles should reflect whole-number ratios.
2. Precision measurements may detect **slight deviations** from expected quantum mass if fractional groupings (as allowed in quantum field theory) are inaccurate.

B. Stable vs. Unstable Particles

1. Stable particles consist of **symmetrically rotating clusters** of Unit Particles.

2. Unstable particles show **rotational imbalances**—this could be modelled and predicted computationally.

16.3 Predictions in Electromagnetism

A. Magnetic Induction Explained by Spin Alignment

1. Magnetic effects are due to **field alignment** of Unit Particles.
2. Induced magnetism in iron occurs as Unit Particles inside align in **rotational harmony**—a claim testable by nano-level field mapping.

B. Field Polarity Shift

1. Magnetic poles should exhibit **reversible spin gradients.**
2. Magnetic reversal, as seen in Earth's geological record, may be modelled as large-scale **field spin oscillation**, not a crustal reorientation.

16.4 Predictions in Gravitational Behaviour

A. Mass-Dependent Attraction

1. Instead of mass alone, **spin field density** determines attraction.
2. Two objects of equal mass but differing spin field configurations might show **gravitational variation**—this is a falsifiable claim.

B. Acceleration Limits

1. UPT implies that **gravitational acceleration** has a ceiling, defined by **spin saturation density** in the field.
2. Black hole-like systems may not be infinitely dense but rather **limit-density constructs** where Unit Particle motion becomes harmonically locked.

16.5 Large-Scale Predictions

A. Cosmic Filament Formation

1. UPT predicts the **web-like structure of the universe** arises from **field-based rotation patterns**, not dark matter scaffolding.
2. Observational surveys (e.g., SDSS) should detect **spin-correlated galaxy alignments**, not random dispersions.

B. Redshift Mechanism

1. Cosmic Redshift may result from **energy loss due to rotational field interaction**, not just space expansion.
2. This suggests **wavelength elongation rates may vary** with field density—offering a new method of distance measurement.

16.6 Simulated Testing and Computational Models

UPT CAN BE VERIFIED or refined using:

1. **Supercomputing models** simulating Unit Particle interactions.
2. **Rotational field simulations** showing how matter evolves from uniform spin matrices.

These models may highlight **natural formation of particles** under UPT rules, helping bridge the theoretical with the observable.

16.7 Challenges and Counterarguments

WHILE UPT IS ELEGANT, it must address:

1. Why haven't spin-density variations been detected in gravitational measurements?
2. How can quantised Unit Particle groups be isolated in high-energy colliders?
3. What accounts for quantum randomness within deterministic field

mechanics?

These challenges guide future refinement of the theory and its interface with current models.

16.8 Conclusion

UNIT PARTICLE THEORY is not merely philosophical—it is scientific, **predictive, and testable**. From particle mass ratios to cosmic filament alignments, UPT provides a roadmap of **specific, measurable outcomes** that can be explored through both laboratory experiments and astronomical observation. If validated, these predictions could lay the foundation for **a new scientific revolution**.

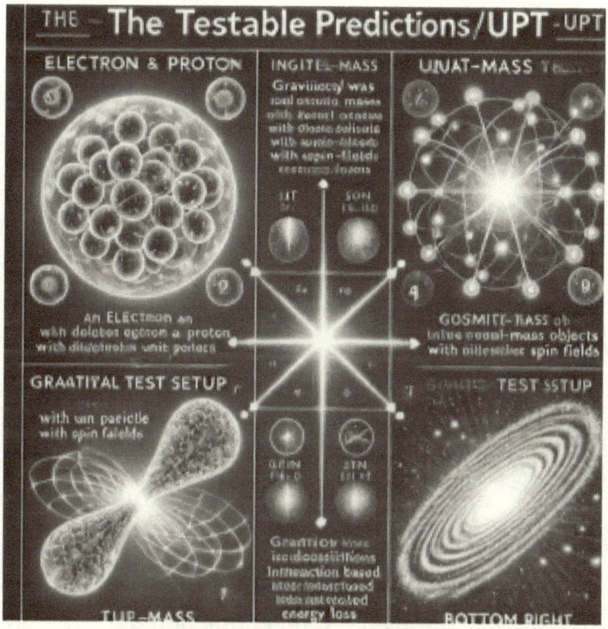

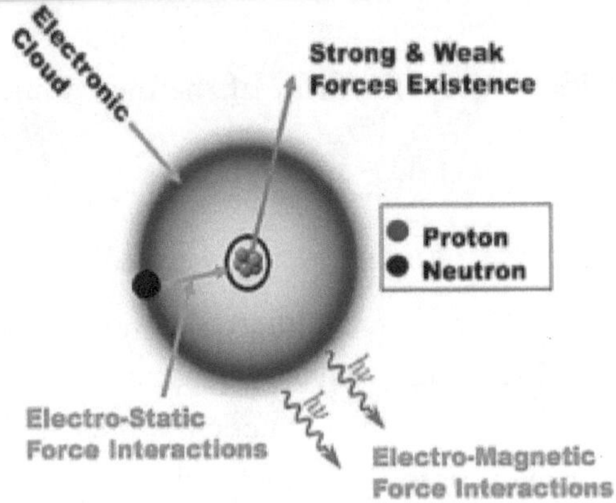

Chapter Seventeen: Conclusion – Toward a Unified Understanding of Reality

17.1 Revisiting the Vision

IN THIS BOOK, WE SET out on a journey to reconsider the **very foundation of the physical universe.** Through the lens of **Unit Particle Theory (UPT)**, we have explored the possibility that **all matter, forces, and cosmic phenomena emerge from a single, indivisible entity**: the **Unit Particle.**

Unlike current theories which describe particles through abstract fields and force carriers, UPT offers a **materialist, rotational model**—simple yet profound. The **Unit Particle** is not only the fundamental constituent of all matter but the cornerstone of every interaction and field, from the gravitational binding of galaxies to the repulsion of magnetic poles.

17.2 What We Have Learned

THROUGHOUT THESE CHAPTERS, we have:

1. Defined the **Unit Particle** as a finite-mass, indivisible entity whose

rotational motion gives rise to all known particles.

2. Shown how **electrons, protons, neutrons**, and even photons emerge from integer groupings of Unit Particles.
3. Reinterpreted **gravity, electromagnetism, and the nuclear forces** as consequences of field gradients and spin dynamics.
4. Proposed a **cosmological model** rooted not in spacetime curvature, but in the field configuration of rotating unit structures.
5. Explored how **dark matter, dark energy, and Redshift** phenomena might be explained without invoking unknown substances or inflationary epochs.
6. Presented **testable predictions**—from mass ratios and magnetic reversals to spin-field Redshift—that could support or falsify UPT.

17.3 Implications for Modern Physics

UNIT PARTICLE THEORY **does not reject modern physics**—it builds upon it by offering **a deeper substrate beneath the quantum and relativistic frameworks.** It proposes that beneath the abstract mathematics lies a physical mechanism: rotation, density, and field structure of real entities.

If verified, UPT could:

1. Replace the **Standard Model's zoo of particles** with a single building block.
2. Supersede **General Relativity's curvature-based model** with a field-density approach.
3. Create **a unified theory** where no force is fundamentally distinct—only varying field conditions and rotational geometries.

17.4 The Road Ahead

THIS THEORY IS IN ITS infancy. Much remains to be done:

1. **Mathematical formulation** of spin density fields.
2. **Laboratory-scale experiments** to detect differential spin effects.

3. **Astronomical surveys** to verify cosmic alignment predictions.
4. **High-resolution simulations** to model formation of matter from Unit Particles.

Most importantly, UPT requires **open-minded investigation**, not dogmatic dismissal. Science must remain fluid, open to paradigms that challenge convention if evidence supports them.

17.5 A New Vision of Reality

IN THE GRAND VISION of Unit Particle Theory, the universe is no longer a disjointed system of forces and random emergence—it is a **symphony of spin**, a cosmic ballet of undividable, eternal particles, shaping everything from atoms to galaxies.

Whether or not UPT proves to be the final theory, it presents a compelling reminder that **the quest for truth is never complete**. It is only through bold ideas, rigorous questioning, and deep curiosity that science advances.

May this theory ignite inquiry, stir imagination, and, perhaps, bring us one step closer to understanding **the true nature of existence**.

Appendices

Appendix A:

Mathematical Representations of Rotational Fields

This appendix introduces simplified mathematical expressions to support the theory of field behaviour driven by rotational dynamics of Unit Particles.

Key Equations:

1. **Angular Momentum Density:**

$$L(r) = m . r^2 . \omega(r)$$

1. **Unit Particle Field Density Gradient:**

$$\rho(r) = \rho 0\left(\frac{r0}{r}\right)$$

n

THESE EXPRESSIONS ILLUSTRATE how the mass and rotational velocity of unit particles influence field strength and structure.

Appendix B: Comparative Table of Forces

Force	Standard Model View	UFT Perspective
Gravity	Warping of spacetime	Rotational density field around massive objects
Electromagnetic	Vector field and wave-particle duality	Alignment of unit-particle spin circuits
Strong Nuclear	Gluon-mediated interaction	Condensed UP field with maximum density
Weak Nuclear	Mediated by W and Z bosons	Transitional alignment shift in unit structures

This table summarises the contrast between conventional physics and UFT's reinterpretation of forces as manifestations of one core mechanism.

Appendix C:
Glossary of Key Terms

1. **Unit Particle (UP):** The indivisible, core constituent of all matter and fields.
2. **Rotational Field:** A spatial arrangement of unit particles spinning around a central axis.
3. **Field Density:** A measure of the number of UPs in a given volume of field space.
4. **Spin Circuit:** A closed rotational pattern enabling force expression, such as magnetism.
5. **Polarity:** The directional influence produced by spin alignment (e.g., magnetic north/south).

Appendix D: Suggested Readings and Resources

To complement this theoretical framework, the following works provide background, parallels, or contrasting views:

1. **Misner, Thorne & Wheeler** – *Gravitation*
2. **Brian Greene** – *The Elegant Universe*
3. **Carlo Rovelli** – *Reality Is Not What It Seems*
4. **Peer-reviewed papers** on:
    - Spin and angular momentum in quantum systems
    - Rotational field models in astrophysics
    - Alternative unification theories

Bibliography

1. **Einstein, A.** (1920). *Relativity: The Special and the General Theory.* New York: Henry Holt and Company.
2. **Feynman, R. P., Leighton, R. B., & Sands, M.** (1963). *The Feynman Lectures on Physics.* Addison-Wesley.
3. **Greene, B.** (1999). *The Elegant Universe: Superstrings, Hidden Dimensions, and the Quest for the Ultimate Theory.* W. W. Norton & Company.
4. **Hawking, S. W.** (1988). *A Brief History of Time.* Bantam Books.
5. **Kuhn, T. S.** (1962). *The Structure of Scientific Revolutions.* University of Chicago Press.
6. **Misner, C. W., Thorne, K. S., & Wheeler, J. A.** (1973). *Gravitation.* W. H. Freeman and Company.
7. **Rovelli, C.** (2016). *Reality Is Not What It Seems: The Journey to Quantum Gravity.* Penguin Books.

8. **Susskind, L.** (2005). *The Cosmic Landscape: String Theory and the Illusion of Intelligent Design*. Little, Brown.
9. **Weinberg, S.** (1977). *The First Three Minutes: A Modern View of the Origin of the Universe*. Basic Books.
10. Articles from *Physical Review D*, *Nature Physics*, and *Scientific American*, accessed between 2020–2024.
11. Selected lectures and research from MIT Open Courseware and CERN publications.
12. **Ghoshal, M.** (2025). *Materialism Theory and the Concept of Unit Particles*. [Self-published].

ABOUT THE AUTHOR

Manoranjan Ghoshal is an independent researcher in physics and a visionary thinker. He is the author of *Materialism Theory and the Concept of Unit Particles* and now expands his revolutionary ideas in *The Unity of Forces.* His work challenges established physics with bold, accessible theories that seek to unify the forces of nature under a single framework.

Manoranjan ghoshal Page

Don't miss out!

Visit the website below and you can sign up to receive emails whenever manoranjan ghoshal publishes a new book. There's no charge and no obligation.

https://books2read.com/r/B-A-SODHD-JAXEG

BOOKS 2 READ

Connecting independent readers to independent writers.

www.ingramcontent.com/pod-product-compliance
Lightning Source LLC
LaVergne TN
LVHW090416210926
848480LV00002B/590
9798231740437